WHAT YOUR COLLEAGUES ARE SAYING . . .

This book will be a game-changer for secondary mathematics teachers. I especially appreciate the inclusion of activities that integrate SEL and UDL and explain how they can be uplifted in instructional practices. Implementation of the practices described will indeed transform math instruction in 6–12 classrooms.

Latrenda Knighten
President, National Council of Teachers of Mathematics (NCTM)
Baton Rouge, LA

Kudos to Michael D. Steele and Joleigh Honey for this hands-on guide! They give us concrete tools for shifting our beliefs, implementing classroom routines that operationalize what we say we believe, and collaborating to re-envision the systems we operate in. Whether you work in one classroom or support teachers in many classrooms, you'll want to include this invaluable resource in your must-read pile.

Cathy Seeley
Past President, NCTM
McDade, TX

Our education lexicon is filled with often overused—and seriously misunderstood—terms like *rigor, differentiation,* and (sadly) *asset-based.* Fortunately, in this accessible, practical, and powerful book, Steele and Honey help us convert "asset-based" into a way of thinking and acting that can truly transform our interactions with students and classes.

Steven Leinwand
Principal Researcher, American Institutes for Research
Washington, DC

Steele and Honey's book on asset-based language, routines, and systems expertly weaves together best practices for mathematics instruction and diverse learners. The care and attention to detail for how each decision we make can be grounded in an asset- or deficit-based perspective is something that can impact the trajectory of all of our students. I can't wait to have my teachers dive into this book, and wish it was something I had early in my career to help me see how each of these components work together to create an asset-based learning environment. I know this book will be a tool used to strengthen instruction and bolster the mathematical identities and futures of students whose teachers choose to leverage its potential.

Melissa Robbins
Director of Mathematics, LEAD Public Schools
Hermitage, TN

Steele and Honey have successfully shown us how we can take the assets-based views of teachers and students advocated by organizations like NCTM, NCSM, ASSM, and AMTE and turn them into practical actions every teacher and leader can take to make our math classrooms more equitable and student-centered.

Paul Gray
Past President, NCSM: Leadership in Mathematics Education
Dallas, TX

This book is a great resource for taking educators through the change process of using asset-based language, routines, and systems in place of the current practice. The inclusion of strategies for a variety of stakeholders (teachers, math coaches, district personnel, state personnel, the community, and more) makes this book a valuable tool that can be used for professional learning community (PLC) work and book studies. Its connections to aspects of social-emotional learning (SEL) and universal design for learning (UDL) math provide added value. This book is a must-have.

Shelly M. Jones
Professor, Connecticut State University
Hamden, CT

I believe in students and will bet on them being capable of grade-level work every time. We should all teach in a way that embraces what students bring with them to our classrooms. This book supports teachers in shifting their practice so that it supports positive student identity through asset-based actions.

Travis Lemon
Teacher, Instructional Coach, Author Consultant, Alpine School District
Lehi, UT

The authors exhibit a strong yet non-judgmental understanding of the unintentional consequences of status-quo deficit-based situations, and gently guide the reader to easy-to-implement alternatives that better support the desired outcomes of all educators: more students finding the joy and beauty of mathematics as they learn content deeply. Read this book now and do what you can. Not only will student outcomes improve, but educators who are tired and burned out will find new joy and energy in this important work.

Joanie Funderburk
Strategic Alliance and State Policy Director, Texas Instruments
Past-President Colorado Council of Teachers of Mathematics Affiliates
Coordinator, NCSM
Littleton, CO

This book makes you think about the messages students get from their teachers, schools, and school system structures. Could you unintentionally be sending students negative messages? Dr. Michael D. Steele and Joleigh Honey support you to reflect and adjust actions and practices at all levels of the systems to make sure all students know they are valued and have potential for mathematical brilliance!

Katherine Arrington
NCSM President
Director of Systemic Transformation,
Charles A. Dana Center at the University of Texas at Austin
Austin, TX

Transform Your Math Class Using Asset-Based Teaching for Grades 6–12 provides practical, applicable strategies that support teachers in keeping students at the forefront of our decision-making. In our dynamic, diverse world, this book provides math teachers with tools that can meaningfully transform their practice and empower *all* students to see themselves as doers of mathematics.

David Dai
Math Instructor, Mobile County Public School System
2022–2025 NCTM Board Director
Mobile, AL

This book is a must-read for anyone who is ready to truly transform their secondary mathematics classroom and leverage the mathematical assets of learners!

Lindsey Henderson
Secondary Mathematics Specialist, Utah State Board of Education
Salt Lake City, UT

Whether you are new to asset-based teaching or you've been doing your best to move in that direction for years, *Transform Your Math Class Using Asset-Based Teaching for Grades 6–12* will help you sharpen your understanding and implementation. It's truly an all-in-one practical toolbox to support and challenge individuals or groups of teachers.

Ted Coe
VP Academic Advocacy, Mathematics, NWEA
Scottsdale, AZ

The authors have done a masterful job of describing how an asset-based perspective can transform a classroom into an environment where students' thinking is elicited and used to advance their mathematical understanding and development of identity. The examples provide vivid illustrations of how the language we use, the routines we employ, and the structures we create make a difference in students opportunities to learn and grow. An eye-opening read!

Margaret (Peg) Smith
Professor, Emerita, University of Pittsburgh
Gibsonia, PA

This book is incredibly thought-provoking and will challenge the reader to carefully examine current practices. There are so many wonderful suggestions of relatively minor—but very significant and impactful—changes that can be made in the secondary classroom to be more asset-focused. The classroom vignettes and examples are especially powerful to help in putting the theory into practice.

Kevin Dykema
President, 2022–2024, NCTM
Math Teacher, Mattawan (MI) Consolidated Schools
Mattawan, MI

This book is a game-changer that equips us with tools to transform our classrooms into an environment where students' assets are prioritized. Steele and Honey invite us to extend our thinking by reflecting on, questioning, and enhancing the common routines and procedures we regularly implement. Both new and seasoned educators will gain valuable next steps to shift current practices and improve mathematics education for all students. The authors share their own experiences and learning journeys, offering tips to support educators with actionable steps toward creating a more asset-based setting that welcomes all learners to succeed.

Kristi Martin
Mathematics Teacher, Tumwater School District
Tumwater, WA

TRANSFORM
YOUR MATH CLASS USING
ASSET-BASED TEACHING
FOR GRADES 6–12

MICHAEL D. STEELE • JOLEIGH HONEY

TRANSFORM
YOUR MATH CLASS USING
ASSET-BASED
TEACHING
FOR GRADES 6–12

CORWIN Mathematics

For information:

Corwin
A SAGE Company
2455 Teller Road
Thousand Oaks, California 91320
(800) 233-9936
www.corwin.com

SAGE Publications Ltd.
1 Oliver's Yard
55 City Road
London, EC1Y 1SP
United Kingdom

SAGE Publications India Pvt. Ltd.
Unit No 323-333, Third Floor,
F-Block
International Trade Tower Nehru
Place
New Delhi – 110 019
India

SAGE Publications Asia-Pacific
Pte. Ltd.
18 Cross Street #10-10/11/12
China Square Central
Singapore 048423

Vice President and Editorial Director:
  Monica Eckman
Associate Director and Publisher,
  STEM: Erin Null
Senior Acquisitions Editor, STEM:
  Debbie Hardin
Senior Editorial Assistant:
  Nyle De Leon
Production Editor: Tori Mirsadjadi
Copy Editor: Sheree Van Vreede
Typesetter: Integra
Proofreader: Dennis Webb
Indexer: Integra
Cover Designer: Scott Van Atta
Marketing Manager:
  Margaret O'Connor

Printed in the United States of America.

Paperback ISBN 978-1-0719-3085-4

This book is printed on acid-free paper.

24 25 26 27 28 10 9 8 7 6 5 4 3 2 1

Contents

 Visit the companion website at
https://qrs.ly/xyfid21
for downloadable resources.

Preface

Teaching mathematics is a complex yet rewarding profession. We develop relationships and cultivate critical thinking among our students. But sometimes it is difficult—not just hard but *difficult* work. The complexity of our profession has become exacerbated since coronavirus of 2019 (COVID-19), and yet, somehow, while the world stopped, we did not. We learned new ways to engage students and found out that not everything we did before COVID-19 was always best for students.

The beginning of the authors' asset-based journey together began at a Conference Board of Mathematical Sciences (CBMS) meeting during the early days of COVID-19 when we were collaborating with a small group of presidents from different national organizations about the assets that so many teachers brought with them to support students during the pandemic. During the course of the next few years, we worked with teachers and other mathematics educators and national organization leaders to recognize the power and impact that asset-based language, routines, and systems have made for so many students and their families.

Thank you in advance for taking this journey with us and for creating space to reflect on ways to transform our math classrooms. The process of creating this book began a few years ago after our initial discussion in 2020, but what you will read in this book spans across many years (from the National Education Association Committee of Ten established in 1892 to research findings, interviews, and observations in 2024). We hope that your experience with this book represents our own in writing it. In the beginning, we saw the power in, but also the need for, stronger asset-based perspectives. Over time, we began to uncover areas of our own practice that were less asset based, even deficit based in some instances, than we may have liked. However, this awareness made it possible to gain new insights and consider ways to shift our work toward a more asset-based perspective. An outcome: Our personal lives have also become more asset based! We are more aware when we hear deficit language (which is different than constructive criticism), and we are more intentional in thinking about how we are expanding the sphere of belonging using asset-based perspectives.

Acknowledgments

None of the work you are about to read would be possible without the collaborations that both authors have had with secondary mathematics teachers during the past four decades. From our work as classroom teachers, curriculum developers, teacher educators, professional developers, and leaders, we have learned so much about the work of effective mathematics teaching and learning from the teachers with whom we've had the privilege of working alongside. The work of transforming classroom practice by leveraging students' mathematical assets is challenging, and the opportunities that we have had to bear witness to teachers doing this work in classrooms with students have transformed our own thinking about effective teaching practice. We're also thankful for our professional communities and associations that have trusted both of us with significant leadership roles, allowing us to build our networks, collaborate with colleagues, and better understand how to leverage student assets in the classroom. And of course, our thanks to Corwin and our editor, Debbie Hardin, for her tremendous support and attention to detail (including deadlines!).

Mike would like to acknowledge his wife, Amanda Cooper Steele, and their three children, Linnea, Robbie, and Gracelynn: My family has graciously stood by while I spent late nights and early mornings in the office behind the computer, not to mention the sometimes too-loud animated video chats with Joleigh! My wife and kids are a constant source of inspiration and joy, especially after long days and nights of writing, work in schools, or days on campus. I'd also like to thank three of my professional organizations, the Association of Mathematics Teacher Educators, the Conference Board of Mathematical Sciences, and the National Council of Teachers of Mathematics. The membership of each of these organizations put significant trust and faith in me as a leader, and it is that very leadership that connected me with Joleigh and led directly to the volume you are about to read. I thank my longtime mentor and friend, Peg Smith, who has been kind and generous over the years with her wisdom and collaboration. Peg's talent for listening to teachers, giving reflective guidance that advances thinking, and writing about research-informed ideas in meaningful and accessible ways has guided my own professional work and writing. I'm thankful for extraordinary school district colleagues like Jen Walton in Noblesville (IN) Schools and Kris Devereaux in

Zionsville (IN) Community Schools who have trusted me to foster long-term collaborations with their districts and teachers. It is in their schools and classrooms that my own thinking about asset-based learning environments has been shaped. And finally, to Kate Johnson and Cai Steele, who serve both as educator colleagues and dear friends (the former) and family (the latter), who have been powerfully influential in pushing my thinking about the assets that students (and teachers) bring to the classroom and how we attend to each and every student's needs.

Joleigh would like to acknowledge her husband, Andrew Marchant, and their three children, Alyssa, Dayne, and Fletcher: They have been supportive and encouraging throughout this process and have taught me many things about asset-based perspectives that have transformed my personal life. My husband has mastered being supportive while reminding me that part of this work is giving myself grace to step away. I would like to thank my colleagues (and second family) from the Association of State Supervisors of Mathematics. I have appreciated the many conversations I have had with this group of extraordinary peers on this topic. I would especially like to thank Mary Pittman, Andrew Byerly, and Becky Unker for their passion and feedback. I would like to thank members and the Executive Board of the Conference Board of Mathematical Sciences for always pushing my thinking. Whether we are discussing the Year of Math (2026!); diversity, equity, and inclusion; or opportunities for students to take courses that better align with their future interests, the conversations always put students first and epitomize asset-based perspectives. This includes bringing awareness to deficit-based perspectives that we may have been blind to before our conversation. I'd also like to thank the National Council of Teachers of Mathematics for their never-ending, ongoing pursuits to further the goals of this organization to support and advocate for mathematics teachers. I'd also like to thank Utah math teachers and my Math Vision Project family. For 30 years, I have felt supported and part of a very strong network of people who deeply care for students and strive for them to have a better future. Thank you for all you have taught me and continue to teach me. I'd also like to thank my friends who also love math education. We spend our time off talking about . . . math education, even when we are engaged in other activities. Thank you for your support and for pushing me to be a better human.

PUBLISHER'S ACKNOWLEDGMENTS

Corwin gratefully acknowledges the contributions of the following reviewers:

Katherine Arrington
NCSM President and Director of Systemic Transformation, Dana Center
Austin, TX

Lisa Ashe
President, ASSM
Wake Forest, NC

Kevin Dykema
Math Teacher, Mattawan (MI) Consolidated Schools
President, 2022–2024, NCTM
Mattawan, MI

Joanie Funderburk
Strategic Alliance and State Policy Director, Texas Instruments
Past-President, Colorado Council of Teachers of Mathematics
Affiliates Coordinator, National Council of Supervisors of Mathematics
Littleton, CO

Paul Gray
Past President, NCSM: Leadership in Mathematics Education
Dallas, TX

Shelly M. Jones
Professor, Connecticut State University
Hamden, CT

Latrenda Knighten
President, 2024–2026, NCTM and Math Curriculum Supervisor
Baton Rouge, LA

Travis Lemon
Teacher, Instructional Coach, Author Consultant, Alpine School District
Lehi, UT

Kirsti Martin
Mathematics Teacher, Tumwater School District
Tumwater, WA

Melissa Robbins
Director of Mathematics, LEAD Public Schools
Hermitage, TN

About the Authors

 **Michael D. Steele** is a professor and chairperson of the Department of Educational Studies in Teachers College at Ball State University. He is a past president of the Association of Mathematics Teacher Educators, current director-at-large of the National Council of Teachers of Mathematics, and editor of the journal *Mathematics Teacher Educator*. A former middle and high school mathematics and science teacher, Dr. Steele has worked with preservice secondary mathematics teachers, practicing teachers, administrators, and doctoral students across the country. He has published several books and research articles focused on supporting mathematics teachers in enacting research-based effective mathematics teaching practices.

Dr. Steele's work focuses on supporting secondary math teachers in developing mathematical knowledge for teaching, integrating content and pedagogy, through teacher preparation and professional development. He is the co-author of NCTM's *Taking Action: Implementing Effective Mathematics Teaching Practice in Grades 6–8*. He is a co-author of several research-based professional development volumes, including The *5 Practices in Practice: Successfully Orchestrating Mathematics Discussions in Your High School Classroom*, *Mathematics Discourse in Secondary Classrooms*, and *We Reason and Prove for All Mathematics*. He directed the National Science Foundation (NSF)-funded Milwaukee Mathematics Teacher Partnership, an initiative focused on microcredential-based teacher professional development and leadership. His research focuses on teacher learning through case-based professional development, and he has been an investigator on several NSF-funded projects focused on teacher learning and development. He also studies the influence of curriculum and policy in high school mathematics, with a focus on Algebra I policy and practice, and is the co-author of *A Quiet Revolution: One District's Story of Radical Curricular Change in Mathematics*, a resource focused on reforming high school mathematics teaching and learning. He works regularly with districts across the country to design and deploy teacher professional development to strengthen effective secondary teaching practice.

Dr. Steele was awarded the inaugural Best Reviewer award for Mathematics Teacher Educator and was author of the 2016 Best Article in *Journal of Research in Leadership Education*. He is an active member of and regular presenter for the National Council of Teachers of Mathematics, the National Council of Supervisors of Mathematics, and the Association of Mathematics Teacher Educators. He reviews regularly for major mathematics education and teacher education journals.

 **Joleigh Honey** is an author and consultant and is in her 30th year as a mathematics educator. She is the immediate past-president of the Association of State Supervisors of Mathematics (ASSM), serves on the Executive Committee of the Conference Board for Mathematical Sciences, and is a current director-at-large of the National Council of Teachers of Mathematics. Joleigh has been a secondary mathematics classroom teacher, academic coach and specialist, PK–12 district- and state-level mathematics supervisor, and the PK–12 STEM coordinator and equity specialist for the state of Utah.

Joleigh is an author of *Open Up Resources High School Math*, a consultant for the Launch Years Initiative through the Charles A. Dana Center, and serves on the STEM Identity working group as a member of ASSM. Over the years, she has worked with teachers, specialists, and state supervisors across the country. She has designed and led professional learning focusing on equity, student engagement, and ensuring all students, including students with disabilities, have access to and success with meaningful course level content.

Joleigh was awarded the 2024 Lifetime Achievement Award by the Utah Council of Teachers of Mathematics and is an active member of and regular presenter for the Association of State Supervisors of Mathematics, National Council of Teachers of Mathematics, and National Council of Supervisors of Mathematics Leadership in Mathematics Education.

CHAPTER 1

Introduction to an Asset-Based Perspective

Welcome! The goals of this introduction are to welcome you to the community, introduce you to the "why" of this work, and explain how this book is organized. Much like we would advocate with our students, we'd like to start out by supporting you in thinking about what ideas and assets you bring to this work. This chapter will ask you to think about what asset-based perspectives mean to you, provide you with some of our thoughts, and give you a variety of frameworks and tools that may be useful to you as you build and grow your sense of asset-based perspectives and how they can support student learning in secondary mathematics.

There are many ways that you can use this book. You can read this book for pleasure or reflection, and we anticipate that you'll have many opportunities to add to and reflect on your own teaching knowledge and practice. We also encourage you to use the online study guide to have a more enriching experience individually or with colleagues, perhaps in a professional learning community (PLC).[1] However you are able to or choose to engage in the work, we welcome you to join the asset-based perspectives community.

 The online study guide for this book is available for download at https://qrs.ly/xyfid21

> *"Do the best you can until you know better. Then when you know better, do better." (Maya Angelou)*

Each chapter in this book will begin with some questions to consider. These questions are designed to guide and scaffold your thinking as you engage with the ideas in each section of text. We'll return to the questions (or similar versions of them) at the close of each chapter when we prompt you to Reflect, Apply, and Transform your thinking and practice.

---

[1] PLCs are defined and operationalized in a wide variety of ways. In some places, they're large umbrellas for smaller, coordinated groups of colleagues that teach a course or grade level. In other places, they encompass a department's worth of math teachers. We use PLC in this book in the most inclusive and expansive sense, accommodating those two examples and the many variations between. Engaging in this text with your PLC colleagues brings value, no matter what the size and shape of your PLC.

Questions to Consider

1. What do asset-based perspectives mean to you?

2. How does the language we use, the instructional routines we implement, and the classroom, school, district, state, and national (federal) systems we work in impact student outcomes?

3. Why are asset-based perspectives critical in mathematics education?

What are some things you do differently because you now "know better"? I (Joleigh) have many experiences of this in both my professional and personal life. Professionally, this is my 30th year in education, and it is almost hard to recognize the person I was when I started on this journey as a mathematics educator. One example (of thousands) of where I do better now is classroom management.

When I started teaching, I remember writing names on the board when students were disruptive and placing check marks by their names if the disruption continued. I had a whole structure to this system. Writing someone's name on the board was a "warning" and then a check meant some form of a consequence, with additional checks adding to the consequence. I never felt comfortable doing this, so I rarely did even though this strategy was my primary form of classroom management during my first couple of years as a teacher. It was also a common practice in my school (when I was a student in high school in South Carolina and during my first couple of years as an educator in Utah). I knew teachers who never wrote names on the board and others who had a list daily. I don't know anyone who does this now and would say that this is an example of when our system learned better classroom management strategies, we did better (Terada, 2020). My classroom management is ongoing. Today, I use assets and strengths to focus on building inclusive relationships, engaging students with meaningful, relevant math contexts, and valuing student contributions that position all students as capable of being doers of mathematics. I continue to learn what to do (be present and pay attention so that I recognize the needs of individuals and the group) and what not to do (react to minor distractions).

I continue to learn better when I reflect on my *decisions*, *self-awareness*, and *self-management*, and I reach out to others to support my growth. These Social Emotional Learning (SEL) strategies (CASEL, 2024) have helped me become more intentional in my work. For example, I am more self-aware of my actions and recognize my strengths and areas I want to improve. My strengths include listening to student ideas, hearing what they do know, and cultivating a classroom environment that embodies a community of learning. I am also aware that I can become anxious with everything that takes place in a classroom and do not always recognize when individual students are having a hard day. I am also not as good as I'd like to be at pausing and celebrating student successes within and outside the classroom. I continually think about responsible decision-making when it comes to pacing a lesson and about attending to ensure all students are positioned as contributors in small group and whole group discussions, as well as about the million other choices we make that impact student outcomes. Over time, I have increasingly come to value the importance of relationships. Students and colleagues have provided many lessons that have transformed this work from a job to a rewarding career.

Many people have contributed to my growth in asset-based perspectives throughout the years. Over time, and through my collaboration with my co-author Mike Steele, I have come to synthesize asset-based perspectives that fall under these three main categories: Language, Routines, and the Systems in which we teach.

ALIGNMENT EXERCISE: WHY ASSET-BASED PERSPECTIVES?

Each chapter of this book includes an Alignment exercise. The purpose is to engage in a reflection or an activity aligned to the chapter. A supplemental study guide aligns with all of the activities included in this book that you can download and use as a resource. The study guide provides you with additional reflective activities that you can engage in by yourself or if you are working with a group of colleagues.

 The online study guide for this book is available for download at https://qrs.ly/xyfid21

In Chapter 1, the Alignment exercise is an introduction to the book. We will consider our current understanding of asset-based perspectives, including what they are and why they are essential. Pause and reflect on the following questions before reading the rest of the chapter.

■ ■ ■ Try This

Reflect on the following questions:

1. What does it mean to have an asset-based perspective?

2. How are asset-based perspectives similar to and different from a strengths-based or growth mindset view? What other education initiatives seem similar (or contrary) to asset-based perspectives?

3. Why is an asset-based perspective essential in education today? ■

We hope this exercise elicited your understanding of asset-based perspectives and how they are similar yet distinctively different from other constructs. Let's unpack each of these questions a bit. In doing this, we are sharing with you our experiences and the input of other experts in the field. The answers to these three questions are our answer to the *why* of writing this book.

WHAT DOES IT MEAN TO HAVE AN ASSET-BASED PERSPECTIVE?

Asset-based perspectives mean starting with what's already there or what is known instead of focusing on what's missing. To start with what's there, we must first learn to listen for the reasoning and sense-making of the person or people we are communicating with. We ask questions or engage in a task that surfaces ideas that engage students in activating background knowledge. This work recognizes that all students bring prior experiences, strengths, talents, and resources to the learning process and can contribute meaningfully in an authentic learning environment (Association of State Supervisors of Mathematics & Association of Mathematics Teacher Educators, 2024). Student thinking is central and is valued. Mathematics learning environments that are asset based feature students and teachers using language that draws on mathematical strengths and teachers using routines designed for students to surface those strengths and build meaningfully on them toward new learning goals. In asset-based learning environments, the teacher facilitates discussions that amplify what students know and aligns and builds those ideas toward the lesson's content goals and learning progressions.

This book will provide multiple aspects of using asset-based perspectives that increase student outcomes through the lens of language, routines, and our systems. In future chapters, we will highlight that specific language, routines, and systems are not strictly asset or deficit based but their use falls somewhere along a deficit-to-asset continuum. Changing our classrooms is not necessarily about discarding old practices and adopting new ones; it can also be about reshaping and revising existing practices to better reflect asset-based perspectives. We'll use illustrations of this deficit-to-asset continuum throughout the book to remind us all that adopting asset-based perspectives is about a continuum, or a movement over time, not about flipping a switch.

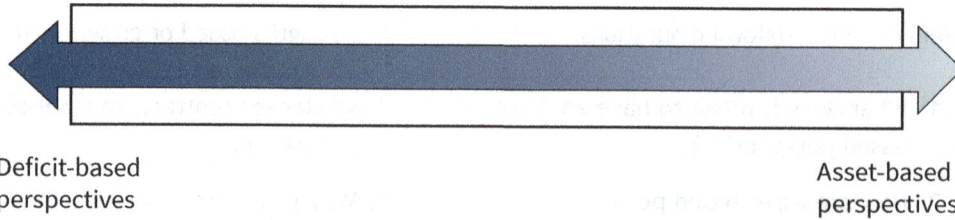

Deficit-based
perspectives

Asset-based
perspectives

Table 1.1 compares key features of asset- and deficit-based perspectives. The distinction is not about who we are as educators but about how we attend to situations. Do we start with what is accurate or good with a situation, or do we start with what is inaccurate or missing? Do we listen and use student thinking, or do we tell steps and say things like "This is easy" or "You just have to follow these steps"? The overview is a first step to reflect on our practices, as well as to do better when we know better. My own efforts are to be more aware when I (Joleigh) am using deficit perspectives and to do my best to shift to be more intentional with implementing asset-based perspectives. As you read Table 1.1, what does the asset-based perspective look like in a classroom? What does the corresponding deficit-based perspective look like? How would an intentional focus on using asset-based perspectives enhance your practice? What would it mean for your classroom and your math department or school?

TABLE 1.1   Characteristics of Asset-Based Perspectives and Deficit-Based Perspectives

ASSET-BASED PERSPECTIVES	DEFICIT-BASED PERSPECTIVES
Center students' current understandings and work	Focus on errors, mistakes, or incomplete thinking as something that needs to be immediately repaired
Consider students' current state of understanding as a foundation on which to build	Center what students *should have done* rather than what students did
Instructional routines are implemented with the mindset that students have lived experiences and funds of knowledge to draw from	Instructional routines are implemented with the assumption that students do not have funds of knowledge to draw from
Provide students with clear feedback about their current performance and how to build on it	Teacher lesson notes perceive deficits or lack of knowledge in student work without a clear path to improve
Teacher voice and actions strengthen students' identities as competent doers of mathematics	Teacher voice and actions foster fixed-ability mindsets and negative mathematics identities
Recognize that students can be successful in any math pathway and that they select courses based on their interests	Label students and place them into tracks based on perceived ability

HOW ARE ASSET-BASED PERSPECTIVES SIMILAR TO AND DIFFERENT FROM A STRENGTHS-BASED OR A GROWTH MINDSET VIEW?

All three of these constructs have much in common. They all impart the value of believing in students and promote the importance of thinking and reasoning as central to deeply learning mathematics. All three, implemented in concert, result in students experiencing joy, beauty, and wonder in mathematics. To truly transform our classrooms, we must recognize the differences between each of these. Let's look more closely to see how each is unique.

Mindset is about how individuals see the world. According to Carol Dweck (2016), individuals who believe their talents can be developed (through hard work, good strategies, and input from others) have a growth mindset. Those who develop and cultivate a growth mindset achieve more because they try more and are less worried about "looking smart" (O'Keefe et al., 2018). They recognize mistakes as opportunities for learning. Conversely, a fixed mindset is when individuals think that talents or interests are inherent in a person. A conception about this work, or education initiative, is that people either have a fixed or a growth mindset. This conception needs a tweak: Everyone is a mixture of fixed and growth mindsets, continually evolving with experience (Dweck, 2016).

A distinction between *asset-based perspectives* and *growth mindset* is that in an asset-based perspective, we always start with what students know. In a growth mindset space, we believe students are capable, but our lens is from the space of considering what is not known. . . yet (Boaler, 2022). This small distinction is quite important. Although both recognize that students can be successful in mathematics at high levels, how we enter the work is different.

Strengths-based perspectives are similar to asset-based perspectives in that they both start with what students know as opposed to what is missing. A strengths-based focus uses an appreciative inquiry model in which we first identify and then focus on using one's strengths to support learning (Kobett & Karp, 2020). Examples of different types of strengths that leverage students funds of knowledge include focusing on quantitative reasoning, structural thinking, and repeated reasoning (Kelemanik et al., 2016) or on disposition, processes and practices, and content (Kobett & Karp, 2020). Asset-based perspectives, such as those described in this book, pay greater attention to how we emphasize student choice, which may or may not align with specific strengths, when engaging in mathematics (more about this later!).

We feel there is great value in both strengths-based and growth mindsets. This book, however, takes a deep look into how asset-based perspectives can transform the thousands of decisions we make each day. We can look at the decisions we make in each teaching moment as having a particular location on the continuum of deficit-to-asset perspectives. These decisions and how we operationalize them impact students and our work in the classroom, as well as across our educational system. These decisions can either resonate with one another or clash, which can result in neutralizing or negating important asset-based efforts. Throughout each chapter, we stop, step back, and pay attention to our beliefs and notice where they do and do not align with our behaviors. We will analyze structures and identify shifts that can transform outcomes for students and teachers. This work recognizes that one's identity and strengths in mathematics are not fixed but change with experiences. We will address methods for building strengths and cultivating a positive mathematics identity, meaning that we position students so that they see themselves, and others see them, as capable (Aguirre et al., 2024).

WHY IS AN ASSET-BASED PERSPECTIVE ESSENTIAL IN EDUCATION TODAY?

We need asset-based perspectives because we live in a world that is different than it has ever been. Our previous systems of schooling, particularly in mathematics, were built with the express purpose of identifying who was equipped for further mathematical study and who was not. This system inherently embodied deficit-based perspectives. These systems have resulted in countless students seeing themselves as incapable of making sense of mathematics and of valuing its role in the world. If we didn't need asset-based perspectives in teaching mathematics, we would have far more students saying that they enjoy math and that they see why learning math is helpful to them in their lives than the minority of students who report so now. We would have far fewer people saying they are not good at math.

I (Joleigh) have interviewed close to 1,000 students in several states over the years about their experiences in math in my role as a state mathematics supervisor, an author, and a consultant. Most students have expressed that they take math either because it is required or because it will help them in their next math class (or to get into college). Students generally do not describe math as something that is currently meaningful to them in their lives. How do we stop the perpetual message that math is only for a few or that math is about getting an answer? We stop the myth that math is only for some and that it is only about computation through asset-based perspectives. We need asset-based perspectives for students, and we need them for ourselves. Asset-based language, routines, and systems have the power to transform our classrooms and how our communities view mathematics.

Asset-based perspectives resonate strongly with the strategies described in the CASEL Social Emotional Learning (SEL) Framework (2024) mentioned earlier in the chapter. Starting with ourselves, we need time to self-manage, make responsible decisions from a larger perspective, reflect on our self-awareness, *increase* our social awareness, and build relationships. Using this framework also allows us to develop these skills in our students while increasing engagement and content knowledge for all of our students. Yes, every student. Therefore, we need to create learning environments to empower each student and learn how to enhance their experiences so that they have access to and success with content.

Asset-based perspectives can shift our work from being overwhelming to being full of joy. Instead of focusing on fixing perceived mistakes or deficiencies one individual at a time, we build a community of learning in which students process concepts, make conjectures, and construct viable arguments. The community is structured using ideas and strategies that enhance learning. The teacher provides a classroom environment that cultivates a community of learning in which all students are positioned as meaningful contributors. Likewise, students are responsible for communicating their reasoning and building their understanding from the community via speaking, reading, writing, and listening. They also extend and refine their thinking, just as mathematicians do. Using asset-based perspectives doesn't mean that we all already have the answers but that we can all reason about mathematical ideas and modify our conceptions as we learn new information.

Our journey will focus on asset-based language, routines, and systems that promote positive mathematics identities and support the learning of meaningful mathematics content. In *Catalyzing Change* (National Council of Teachers of Mathematics [NCTM], 2018), the authors discussed the importance of a positive mathematical identity and defined mathematical identity as:

> *the way in which people think of themselves in relation to mathematics: Having a positive mathematical identity means that people feel empowered by mathematics and as doers of mathematics, see the multiple purposes for learning mathematics, appreciate why mathematics is important in their lives, and come to believe that they can succeed in mathematics. (p. 25)*

ORGANIZATION OF THIS BOOK

Throughout our research and discussions with fellow mathematics educators, it has become clear that language, routines, and systems implementing asset-based perspectives can elevate student outcomes and increase morale among all involved in mathematics learning and teaching.

PART 1: ASSET-BASED LANGUAGE

In Part 1 of this book, we address asset-based language. The language in our classroom, school, and the larger system all contribute to our perceptions about mathematics and one's abilities. The first section of this book takes a closer look at the norms in our classrooms that communicate our beliefs, how we attend to the development of academic language, and the conversations we have about mathematics and our beliefs about the abilities of our students.

- Chapter 2: Mathematics content and asset-based language intersect. We provide examples of how language can be both mathematically meaningful and precise and build on and honor the language assets students bring to the classroom.

- Chapter 3: Specific teaching moves can be used to build on students' language assets in the moment-to-moment work of teaching. We focus on these in this chapter.

- Chapter 4: Analysis of our language choices and how they evolve and change over time. We consider the language that we use to discuss specific groups of students and ways to describe student groups and students performance that focus on student assets.

PART 2: ASSET-BASED ROUTINES

In Part 2 of this book, we address asset-based routines. Our classrooms' routines include those that are structural, those that are instructional, and the combinations of certain routines that become practices.

- Chapter 5: Structural routines guide our daily decisions in our classrooms. They are the what, how, and why of what we do. Examples of structural routines include warm-ups, exit tickets, instructional routines, and homework (or checks for understanding). This chapter will unpack aspects of our structural routines that cultivate asset-based learning environments and aspects of structural routines that need disrupting or that fall on the deficit side of the deficit-to-asset continuum.

- Chapter 6: Instructional routines are the decisions we make about implementing instruction. This chapter focuses on instructional routines that continually support students in developing a positive mathematics identity. Instructional routines are not simply asset or deficit based. Learning to implement routines as intended takes an asset-based routine and moves it further along the continuum.

- Chapter 7: After considering various instructional routines, this chapter will deepen our awareness of how to put asset-based instructional routines together in a coherent structure to establish asset-based practices. We highlight the Five Practices for Orchestrating Productive Discussions as an example.

PART 3: ASSET-BASED SYSTEMS

In Part 3 of this book, we address asset-based systems. Many moving cogs go into the system of educating children, from the structures in our classroom to the laws and policies at the state and federal levels. We will examine the many structures that impact our work.

- Chapter 8: Classroom teachers and instructional leaders make thousands of decisions every hour. Some decisions are in the moment, whereas others are more overarching and create the structures of our school and classrooms. This chapter will increase our awareness of how these structures contribute to asset-based learning environments. In addition, we will provide suggestions and ideas to disrupt aspects of our structure that do not benefit students or teachers.

- Chapter 9: Academic coaches, district specialists, district personnel, state-wide personnel, and federal employees also impact the structure of the work we do. This chapter will support current and future employees who work at these levels to stop, reflect, analyze, question, and take action to enhance asset-based structures and disrupt those less asset based.

- Chapter 10: This chapter draws across chapters 8 and 9 and provides suggestions for how to start conversations to foster more asset-based systems in your school and district.

FORMAT OF EACH CHAPTER

Each chapter starts with a short activity aligned with the chapter's theme and is designed to elicit the assets you bring to the chapter's ideas. The Alignment exercises may be a reflection on a current practice, reading a vignette and analyzing information, unpacking routines, and so on. These activities are written to take approximately 15 minutes. We recommend spending time on this before reading the rest of the chapter. The accompanying study guide provides space to capture questions to consider, complete the Alignment exercises, reflect and discuss ways to transform our practice, and activities that provide learning communities with opportunities to deepen thinking about the topic and its connection to your local practices and contexts. If you are reading this book as part of a PLC, we also encourage you to discuss the Alignment exercise with colleagues. The intention is to encourage discourse and contribute to each other's understanding of your diverse perspectives.

Each chapter continues with an exploration of an aspect of mathematics classroom language, routines, or systems from an asset-based perspective. This journey makes use of classroom and school vignettes and provides tips that can be used immediately in the secondary mathematics classroom. We also make connections to three frameworks throughout the book that can support the development of more asset-based learning environments: Social Emotional Learning (SEL), Universal Design for Learning (UDL), Special Populations, and NCTM's Effective Teaching Practices.

The Digging Deeper feature in each chapter is a spotlight on a particular population that showcases how the asset-based strategies support the population being spotlighted. Although the content shared in the chapter is important for all students, the spotlight explicitly calls out how and why it makes a difference to specific populations.

The Reflect, Apply, Transform section returns to the questions at the start of the chapter to support reflection and action items that move your work toward more asset-based perspectives and have a strong impact for students and teachers.

THEMES THROUGHOUT THE BOOK

As we worked on writing this book, we noticed patterns or themes that kept coming up in each chapter in the form of frameworks for the teaching and learning of mathematics that may be familiar to you. We have chosen to embrace these and have identified them using icons. By making these connections explicit, we invite you as readers to make connections to learning that you have already undertaken related to these themes to accelerate your classroom practice toward more asset-based perspectives.

The themes you will encounter are as follows:

- Social Emotional Learning (SEL)

- Universal Design for Learning (UDL)

- Student Populations

- NCTM's Effective Teaching Practices

Look for the icons throughout the book to indicate discussions especially related to these themes.

Social Emotional Learning

Universal Design for Learning

Student Populations

NCTM's Effective Teaching Practices

SOCIAL EMOTIONAL LEARNING (SEL)

The CASEL Framework (2024) addresses five areas to advance student learning and development. The five areas are self-management, self-awareness, social awareness, relationship skills, and responsible decision-making. An overview of each is shown in Table 1.2.

TABLE 1.2   The CASEL Framework

SELF-MANAGEMENT	SELF-AWARENESS	SOCIAL AWARENESS	RELATIONSHIP SKILLS	RESPONSIBLE DECISION-MAKING
Manage one's emotions, thoughts, and behaviors effectively in different situations and to achieve goals and aspirations	Understand one's own emotions, thoughts, and values and how they influence behavior across contexts	Understand the perspectives of and empathize with others, including those from diverse backgrounds, cultures, and contexts	Establish and maintain healthy and supportive relationships and effectively navigate settings with diverse individuals and groups	Make caring and constructive choices about personal behavior and social interactions across diverse situations

SOURCE: Adapted from CASEL.org (2024).

When you see the SEL icon, we encourage you to return to this table to identify the facets of the CASEL Framework that are at play in the discussion. You might reflect on how the language, routine, or system being discussed interacts with the five categories in the CASEL Framework and how the categories can support stronger asset-based perspectives in the mathematics classroom.

UNIVERSAL DESIGN FOR LEARNING (UDL)

The Universal Design for Learning (UDL) framework is based on scientific insights into how humans learn (CAST, 2018). The framework (see Figure 1.1) includes guidelines with descriptions on how to provide multiple means of engagement, representation, and action and expression. Lambert (2021, 2024) adapted this framework with specific attention to the learning of mathematics through the design elements of engagement (meaningful mathematics and supportive classroom environment), representation (focus on core ideas and multimodal), and strategic action (understanding self as a math learner and equitable feedback).

FIGURE 1.1 UDL Math Design Elements

Do your students feel safe enough to take mathematical risks? Are they building relationships in and through math?

Supportive Classroom Environment

ENGAGEMENT

Is the math meaningful and relevant to students? Do students regularly engage in sense-making?

Meaningful Mathematics

Is math content accessible? Multimodal? Can students choose how they solve problems? Are representations connected to each other? To concepts?

Multimodal

REPRESENTATION

Our Goal: Strategic Sense-Makers

Invest Time in Core Ideas

Does the design of instruction guide students to understand core mathematical ideas? Mathematical representations? Develop strategies?

What do your students learn about themselves as math learners? How do you support strategic development?

Understanding Self as a Math Learner

Equitable Feedback

Does feedback help students grow as mathematicians? Is assessment equitable for all learners?

STRATEGIC ACTION

SOURCE: Reprinted with permission from Lambert (2024).

When you see the UDL Math icon, you might return to this framework and consider which of the design elements relate to the language, routines, or systems being discussed. How might your practice better incorporate aspects of these design elements to promote a more asset-based mathematics classroom learning environment?

STUDENT POPULATIONS

As noted, sometimes we want to highlight student populations that deserve additional consideration. Aspects of our system were created years ago that created structures we still use today in our classrooms. Some of these structures must be disrupted if we adapt asset-based perspectives. Some of the populations highlighted include:

- Students who are multilanguage learners

- Students who have mathematics anxiety

- Students with disabilities, including neurodiverse students, or students identified via the Individuals with Disabilities Education Act (IDEA) of 1990 (reauthorized 2004)

- Students who are identified as having a propensity for mathematics

- Students who are Black, Indigenous, and people of color (BIPOC), including those who identify as Hispanic, Pacific Islander, or other racial/ethnic groups you serve. Throughout this book, we will identify specific populations of students where appropriate and use BIPOC when referring to students of color.

- Teachers

- Other community members

When you see the Student Populations icon, you might consider how students similar to the students being discussed are afforded opportunities to demonstrate their assets in your mathematics classroom or the classrooms of teachers you support. How might we provide more explicit and clear opportunities for different student populations to identify and leverage their assets? And how might we be attentive to the deficit-based messages these groups may have been exposed to in the past and shift that thinking?

NCTM'S EFFECTIVE TEACHING PRACTICES

The NCTM introduced the Effective Teaching Practices in the 2014 publication of *Principles to Actions: Ensuring Mathematical Success for All*. These eight practices represent a synthesis of over three decades of research into effective mathematics teaching, and it's likely you're using some, if not all, of these practices already in your classroom. When you see the NCTM Effective Teaching Practices icon, at least one of the following practices will be integral to the discussion:

- Establish mathematics goals to focus learning.

- Implement tasks that promote reasoning and problem solving.

- Use and connect mathematical representations.

- Facilitate meaningful mathematical discourse.

- Pose purposeful questions.

- Build procedural fluency from conceptual understanding.

- Support productive struggle in learning mathematics.

- Elicit and use evidence of student thinking.

(NCTM, 2014)

When you see the Effective Teaching Practices icon, you might reflect on how the practice or practices being discussed are used in your classroom. You might further consider how one might shift the implementation of the practice(s) toward more asset-based perspectives.

■ ■ ■ Digging Deeper

The Digging Deeper component in each chapter highlights specific student populations as it relates to the chapter. Although many aspects of asset-based perspectives positively impact all populations, we will spend time highlighting how this work specifically has significant impact for different populations of students. Don't worry if your reflection does not include information for each population or only includes a little bit. You will have new insights to add as you read and reflect from chapter to chapter. Feel free to add student populations to this list that may be more specific or relevant to you. For example, if your work includes a group of students with disabilities not included below, add this group. For our introductory chapter, we encourage you to use Table 1.3 to reflect on how asset-based perspectives benefit specific student populations.

TABLE 1.3   Reflections on How Asset-Based Perspectives Benefit Specific Populations

STUDENT POPULATION	HOW WOULD THIS POPULATION SPECIFICALLY BENEFIT FROM ASSET-BASED PERSPECTIVES (LANGUAGE, ROUTINES, AND SYSTEM)?
Students who are multilanguage learners	
Students who have mathematics anxiety	

TABLE 1.3 (*Continued*)

STUDENT POPULATION	HOW WOULD THIS POPULATION SPECIFICALLY BENEFIT FROM ASSET-BASED PERSPECTIVES (LANGUAGE, ROUTINES, AND SYSTEM)?
Students with disabilities	
Students who are BIPOC	
Students who are identified as having a propensity for mathematics	

■

Reflect, Apply, Transform

Asset-based language, routines, and systems can transform our classrooms and our mathematics education community. We will continue discussing the three questions introduced at the beginning of this chapter (Questions to Consider) throughout the book. Now that you have read the Introduction, we ask you to reflect on the questions to consider for the chapter:

1. What do asset-based perspectives mean to you?

2. How does the language we use, the instructional routines we implement, and the classroom, school, district, state, and national (federal) systems we work in impact student outcomes?

3. Why are asset-based perspectives critical in mathematics education?

PART 1

Asset-Based Language

Take in the examples that follow of a teacher-student exchange about a math problem, a teacher launching a task with their class and invoking classroom norms, and two teachers talking about instructional approaches to the unit circle. Based on each of these excerpts, what beliefs might you infer that the teacher holds about the following:

- Students and their competence?

- What the work of knowing and doing mathematics is?

- What it takes to be successful in school mathematics?

EXCHANGE A	EXCHANGE B	EXCHANGE C
S: I'm stuck. I know I need to divide $1\frac{1}{2} \div \frac{1}{3}$ but I don't remember how. T: Remember Keep, Change, Flip? S: Which one do I flip again? T: The second fraction. S: Do I need to change the mixed number? T: Yes. Make it an improper fraction.	T: Let's start the way we usually do: Take 2 minutes to read the problem. Write down what you notice and what you wonder about, and we'll share some noticings and wonderings after everyone has had a chance to think and jot ideas down.	Teacher A: We had a great discussion today with my Honors class about the unit circle and how it relates to trigonometric functions. Teacher B: My class always struggles with that. I give them a handout with the important ratios on the unit circle and we have a quiz the next day to see if they remember. The regular class just can't handle that kind of discussion.

The beliefs that we hold influence our work as teachers, including how we structure our class time, the mathematical ideas that we center in our classrooms, and the language we use with students and our teacher colleagues about mathematics. Thinking about our beliefs and how they influence our classroom practice is important, but it's also important to acknowledge that beliefs are complicated, messy, and sometimes contradictory.

In this part of the book, we consider the language we use in teaching mathematics. As we examine language from multiple perspectives, consider how language can both reflect and influence our beliefs as teachers, the beliefs of our colleagues, and the beliefs of (and in) our students.

Honoring Student Language While Building Meaningful and Mathematically Accurate Content

One of the most important functions of the language we use in the classroom is to convey mathematical meaning. The goal of this chapter is to consider the interactions between mathematics content and language. We will explore how, as teachers, we can take an asset-based approach to language that is mathematically accurate and appropriate. By the end of this chapter, you will have some strategies you can use to elicit and use asset-based language from students, as well as to disrupt deficit-based language in your classroom. Specifically, we consider how our language use can serve to advance mathematical understandings while leveraging student assets.

> *"Students whose whole selves are welcomed in class have more bandwidth to focus on learning." (Ruef, 2020)*

Questions to Consider

1. When you think about students' academic language in mathematics, what ideas come to mind?

2. How do you respond when students use informal mathematics language?

3. What contexts do you use when teaching using real-world mathematics tasks?

4. How does students' language use relate to how we give feedback to students and provide assessment opportunities?

ALIGNMENT EXERCISE: BIG ROCKS

SOURCE: istock.com/joshuaraineyphotography

Teaching is often like our commute to work. Sometimes our commutes are exactly as we pictured them when we got in the car. Other times, we experience events like construction, traffic, weather, or other unexpected acts that change the ways we go from point A to point B. We're used to being flexible in teaching, especially with the hundreds of small changes and adjustments we make in a class period. We want to focus here on the *big rocks* that might happen during teaching—significant challenges that have the potential to stop or derail a discussion within a lesson.

Two brief excerpts follow that describe important mathematics content ideas at the middle and high school levels. Read one or both of the excerpts; you might consider one that's closest to the grade level with which you are most familiar and one that reflects mathematics that you may not have visited in a while. As you read, think about what the big rock might be in each situation and what the teacher might do to navigate around it in a productive way.

The class had been working in small groups on the Cookie Conundrum task for about 10 minutes. Ms. Dohm calls the class together to discuss solutions.

Ms. Dohm: So the situation is that we're putting together boxes of cookies for a party. The party colors are green and silver, so the organizers want to make bags with 5 green cookies and 3 silver cookies each.

We have 200 green cookies. How many silver cookies do we need? Who would like to start us off? *Several hands go up to volunteer.* Okay, Brette?

Brette: Well, first we thought it was just 2 less, so 198. But that didn't make sense. We started thinking about making the bags and how we'd have a whole lot of silver ones left over.

Ms. Dohm: Okay, so what did you do next?

Brette: We started drawing pictures of the bags of cookies and counting.

Andre: Brette, we did the same thing but then we made a table.

Ms. Dohm: Can you show us that table, Andre?

G	5	10	15	20	25	30	35	40	45	50
S	3	6	9	12	15	18	21	24	27	30

Andre: From there, we knew that we could double 50 to get 100, and double 100 to get 200, so we came out with 120 silver.

Ms. Dohm: Okay, let's back up. How did you make each row of your table?

Justice: We just added the same thing each time.

Ms. Dohm: Why was that a strategy that made sense to you?

The group all look at one another; nobody verbally responds initially.

Brette: Well if you just look at my picture . . .

Ms. Dohm: We'll take a look at the picture in a moment. But first I want to hear from the group about their strategy.

A few more seconds pass with no response from the group.

Lawrence (who is not in the group): Just say *proportional.* The columns in the table have to be proportional. (Lawrence gestures at the note on the board that says, *Lesson 7.2: Proportional relationships.*) There, done. Can we go to the next question?

Mr. Mutford's Algebra I class is starting a unit on exponential functions. To get them started, he gave them a task centered on the familiar Pay It Forward idea. He framed the task as follows:

Mr. Mutford is going to try to change the world this summer while we are on break. On the last day of school, he is going to do three good deeds for people. In return, he will ask them to do 3 good deeds for others on the first day of break, and pay those deeds forward the next day. If everyone does as they are asked, how many good deeds will be done on the fifth day of break?

Students are sharing their responses to the task after small group work. Some drew pictures, others made tables, and a few others wrote equations. Mr. Mutford had made note of the strategies students used and made specific choices about whom to call on in the dialogue that follows.

| Mr. Mutford | (after some discussion): So we seem to be gravitating toward 243 as our answer. Maya, would you please share how you thought about it with your diagram? |

(Maya shows a tree diagram in which each person does three good deeds on the next day.)

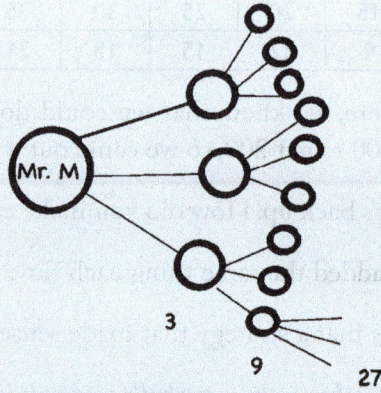

Maya:	Here's how we started. So here's you on the last day of school. You did three good deeds. One of them was for me, just sayin' Mr. M. Then the second day, me and the other two people do three more. Then those people do three more. We got 3 times 3 times 3 times 3 times 3 and that's 243.
Mr. Mutford:	Great! Who has questions for Maya? Janelle?
Janelle:	Can I be one of the other people?
Maya:	I got you. (Maya labels one of the stage 1 figures with Janelle's name.) Better make a list, Mr. M.
Mr. Mutford:	So I don't hear questions about her math. Did everyone do it the same way then?
Panch:	We used exponents, but we got the same thing, 243. Three to the fifth, which is 3 five times.

Mr. Mutford: I heard Panch and Maya both say "times" when they were talking about getting to their answer. But Maya also described each day as being three *more* deeds. Maya, did I get that right? (Maya nods yes.) We usually use *more* to talk about addition. So why are we multiplying here? (*There's no initial response but some murmuring in the room.*) Talk about that for two minutes in your small groups.

(*While groups discuss, Mr. Mutford writes 3 + 3 + 3 + 3 + 3 = 15 on the board next to Maya's diagram.*)

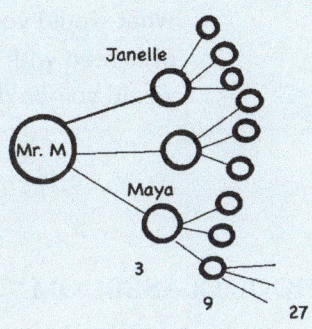

3 + 3 + 3 + 3 + 3 = 15

Mr. Mutford: Okay, so I think what I just wrote is the right answer. Change my mind.

Jay: Nah man, that's not it. You have to multiply. Because it's like each person does three more the next day.

Mr. Mutford: Yes—three *more*. That's addition. Still not convinced. Ramie?

Ramie: Every person does three more, but like the multiplication comes from the days. You have three times as many people doing the deeds from one day to the next. (Ramie begins pointing at Maya's tree diagram.) See like Mr. M is the only one in that first one, then it's like Maya and Janelle and I'mma put me as that third one on the first day of break. Then each of us get three more—shoot, I'm trying not to say more because I know you're gonna roast me, we get three people each to do another one and then it goes from there. It's that each that gets us to the times.

Mr. Mutford: There's a lot of heads nodding here so maybe we're on to something. But riddle me this, I thought Panch used exponents. So are we supposed to add, multiply, or use exponents here to find out how many deeds on Day 5?

Janelle: Oh come on Mr. M, we all got it **right** already.

Discuss these questions with your professional learning community (PLC) or other colleagues. Once you've done so, continue reading and we'll share some thoughts of our own related to these questions.

In the vignette that you read . . .

- What is the math content being discussed?

- What pieces of language (from students or the teacher) seem to be supporting students in understanding the content?

- What pieces of language seem to be getting in the way or causing confusion?

- What would you do next (or if you disagreed with a teacher's choice, what would you do differently)? ■

MS. DOHM'S MIDDLE SCHOOL CLASSROOM

The lesson in Ms. Dohm's classroom is focused on proportional reasoning using a straightforward contextual situation. One can infer that the goal of this lesson is to establish a conceptual basis for the notion of a multiplicative relationship that maintains a common ratio. A common thought pattern for students at this stage is to attend to the additive difference between two quantities, which does not attend to the common ratio. Ms. Dohm's questioning has the goal of teasing out why this relationship is multiplicative rather than additive. While using waiting strategies to afford the group an opportunity to think and respond, a student short-circuits the discussion by making use of the term *proportional* (gleaned from the lesson title) and suggesting that just using this term will suffice as the understanding Ms. Dohm was looking for. This move has the potential to put a halt to students sharing their thinking and grappling with the important mathematical ideas and sends a message that a single answer (in this case, the word *proportional*) is the important part of the mathematical exercise. What might Ms. Dohm do in response to recalibrate after Lawrence's comment?

MR. MUTFORD'S HIGH SCHOOL CLASSROOM

Similarly, Mr. Mutford and his class are grappling with the language for describing exponential relationships in which the rate of change from one stage to the next results in multiplying by a set factor rather than adding. This particular piece of content can be very challenging for students as many of the words in English to describe increases are associated with additive changes. You hear Mr. Mutford pressing on the use of terms like *more* and *times* that tend to have specific associations with arithmetic operations (adding and multiplying). It's also notable that the mathematically precise language of exponents comes out early in the conversation. Mr. Mutford doesn't grab on to that language in that moment; instead, he continues to press for clarification and only brings the term *exponent* back at the very close of the vignette. Janelle's comment at the end expresses what we might hope is good-natured frustration with being pressed to explain, but there's also another implicit message here that the right answer is what's important. What would you do in Mr. Mutford's shoes at this point? Would you let that comment go and move on? Would you say something about it?

* * *

We called this alignment activity *Big Rocks* because in different ways, the mathematical language and ideas expressed at a key moment in each vignette can represent a big rock in language. These pieces of language can serve to close off continued conversation and potentially interrupt or stop entirely the sense-making work that students are doing. In the middle grades vignette, the use of the term *proportional* serves as the big rock (and the student using it does so explicitly to bring an end to conversation). The term *exponent* plays the role of the big rock in the high school excerpt.

Classroom discourse can move quickly, and as a teacher, we're often managing competing demands in the classroom. Some students want feedback or are asking for help, we're checking in on some students to make sure they're making progress and are not silently stuck, and we're always thinking about the next thing we'd like to unfold in the lesson. It's not always clear in the moment when a big rock has been dropped into the road that our lesson is traveling down. And sometimes what may be a big rock to one person may not be to another, especially for us as teachers since we've been thinking deeply about the math ideas in the lesson. Think again about the end of the story about Ms. Dohm. Can you imagine a teacher who heard Lawrence's comment and thought to themselves, *Thank goodness, I was hoping for that word to come out; now let's move on to some notes*? Or maybe you've even had that thought yourself? (We know that we have!) The goal of this Alignment exercise is to help us look out for big rocks in academic language and think about how we can work with those big rocks in asset-based ways.

Now that we've set the stage, in this chapter, we're going to explore together the role of math content in asset-based language. When you think about asset-based language, you may think more about how we elicit and build on the math ideas that students create to focus on their strengths. How might the math content ideas like decimal points, proportional relationships, or exponential functions be asset or deficit based? These ideas may *seem* neutral, but the way that we and our students use the language of mathematics in classroom talk can position these pieces of language in different places on the deficit-to-asset continuum. Knowing the content goal and thinking deeply about the mathematics behind your goal can help us leverage our own language and to think about what to pursue in classroom discourse and what we let go by.

WHAT'S THE ROLE MATH CONTENT PLAYS IN FOSTERING ASSET-BASED LANGUAGE?

As we saw in the opening of this section, the language that we use as teachers and the language students use in talking with us and their peers can foster more asset- or deficit-based perspectives in our classrooms. The words we use, how we use them, and when we use them can position students as capable of knowing and doing mathematics. The what, how, and when of language can potentially make students feel like they should have known a particular idea, that they've described a mathematical idea in an incorrect or improper way, or that they should have remembered something in the moment that they did not. All of these examples can result in students developing a deficit-based perspective.

And our responsibilities as teachers goes beyond the language that we personally might use. Students might use mathematical language in ways that can foster deficit perspectives. This might be intentional at times, such as when a student might correct another with a more mathematical term to appear smart or exert interpersonal leverage. But more commonly these experiences are likely to be unintentional. In using mathematical language that is natural to them but may not be shared by others, a student can unintentionally create a *big rock* situation that shuts down further engagement. As teachers, we need to be thoughtful about how to navigate, and sometimes even disrupt, mathematical language that surfaces in conversation to ensure that each and every learner sees a path forward for themselves to develop new understandings.

Let's look at asset-based mathematical language in the following three ways:

1. Let's think about how we as teachers can position math vocabulary and terminology in our classrooms. This includes the situations directly under our control (*how* we use math terms and *when*) and those that are not (how we respond when students do—and do not—use mathematical language).

2. Let's think about how we handle language when math ideas get more complicated and the terms become increasingly dense and complex. This generally happens as we move into middle and high school, but there are lessons about this idea for every grade level.

3. We'll bring these first two ideas together to help us think about specific teaching actions that allow students to leverage assets. The end goal will be to think about what we want students to walk away with in terms of mathematical language and how we can best facilitate that goal with understanding behind that language.

Throughout this chapter, we're going to pay special attention to multilanguage learners as this population has a particular intersection with the idea of developing academic language. The approaches and tips we share here will benefit all learners, but we also highlight ways in which strategies will be particularly helpful for multilanguage learners. We also provide some additional discussion in the Digging Deeper section. Math terms and math talk have special implications for these groups, and we'll think together about particular actions that work well to support their learning.

But first, a word about our use of language going forward. Already in this chapter, we've talked about math content words and phrases as:

- vocabulary,

- terminology, and

- academic (or mathematical) language.

Sometimes we use these flexibly and interchangeably in our everyday talk, and that's fine! For the purposes of this chapter and book, we'd like to share the key distinctions between these ideas so that *we* have a shared understanding and the terms can help us focus on aspects of language in our classroom and how they can support us in leveraging assets (see Table 2.1).

TABLE 2.1 Understanding the Words We Use to Talk About Math

MATH CONTENT WORDS	DEFINITION	EXAMPLE OF USAGE
Vocabulary	the **words** used in a particular context	We frequently use vocabulary to connote unfamiliar or new words.
Terminology	terms with a **specific technical meaning** in a field of study	This is an important distinction in math—we have some words specific to the domain, like *hypotenuse*, and others that have both mathematical and nonmathematical meanings, like *function*.
Academic language	the **language** needed by students **to be successful in schools**, including oral, written, gestural, and visual representations and the customs and norms of a discipline (Halliday & Webster, 2003)	Vocabulary and terminology are a part of academic language, but it's so much more! In mathematics, it includes the grammar and syntax we use to communicate about math and the culture of how mathematics is discussed and represented. It's important to note that gesture and visuals are important parts of academic language, too.

When we discuss language in this chapter and throughout the book, we'll most commonly be referring to *academic language*. This term is inclusive of both the particular words we use and how and when we use them. The broader context and the explicit inclusion of gestures and representations will help us think as broadly as possible about asset-based approaches to language. Academic language, as you notice, is also inclusive of vocabulary and terminology. When we use those words, we'll be using them in the specific ways they're defined above. We know these two ideas are important too—students need to learn how to use math-specific terminology and vocabulary, particularly for words that have meanings in both colloquial language and math that may be different.

POSITIONING VOCABULARY AND TERMINOLOGY WITHIN OUR CLASSROOM'S ACADEMIC LANGUAGE

Take a moment and think about what your own school mathematics learning experience was like and about how and when vocabulary was introduced and used. How did you learn a new piece of vocabulary or math terminology? How was that new word or phrase integrated into the class's academic language? What were the expectations around how you would use those terms going forward?

If your experience was anything like mine (Mike), the introduction to vocabulary might have looked like what we see in Figure 2.1.

FIGURE 2.1   A Student's Notes From Math Class

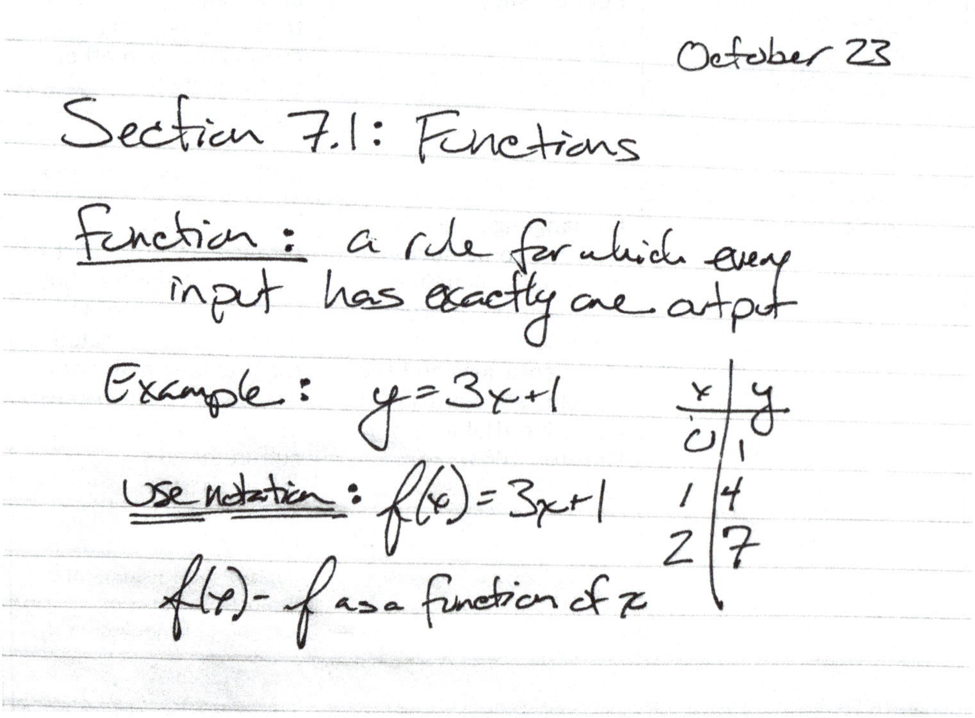

October 23

Section 7.1: Functions

Function: a rule for which every input has exactly one output

Example: $y = 3x + 1$

x	y
0	1
1	4
2	7

Use notation: $f(x) = 3x + 1$

$f(x) - f$ as a function of x

For many students, the start of a unit of study on a new math idea began with a definition of vocabulary written at the top of a textbook page, followed by an example or two. In the discourse that followed, you were expected to make use of that vocabulary when describing the idea.

Consider the notes artifact in Figure 2.1 and the description. As a student, where were there opportunities for me to share my ideas that might relate to functions? Where are my assets and my voice represented? Defining vocabulary and asking students to write down the definition and use the word going forward are not, in and of themselves, deficit-based practices. In fact, they are done with the very best of intentions: To advance their mathematical understandings, students need to know and understand the language of mathematics and use that language in accurate and appropriate ways. Defining a vocabulary word and using that word in examples is certainly one way to do that. But consider whether there are other ways in which as teachers, we can better integrate our treatment of *vocabulary* into the broader sphere of *academic language* that students will use going forward. And consider how we might think about making use of students' assets as we go about introducing new math concepts and the language associated with them.

GETTING READY FOR A POOL PARTY

Let's look at a task from the Algebra 1 or Math 1 (Grade 9) Open Up Resources (2021) curriculum to help us think about vocabulary, terminology, and academic language.

■ ■ ■ Try This

Work on the Example Task, Getting Ready for a Pool Party, a little bit on your own. Then discuss the task with your colleagues. As you do, think about the math content ideas this task addresses and the academic language you imagine students would use in their responses to and discussions about the task. ■

Sylvia has a small pool full of water that needs to be emptied and cleaned, then refilled for a pool party. During the process of getting the pool ready, Sylvia did all of the following activities, each during a different time interval.

Removed water with a single bucket	Filled the pool with a hose (same rate as emptying pool)	Drained water with a hose (same rate as filling pool)
Cleaned the empty pool	Sylvia and her two friends removed water with her three buckets	Took a break

1. Create a story of Sylvia's process for emptying, cleaning, and filling the pool. Number the activities given 1–6 to indicate the order in which they occurred in your story.

2. Sketch a possible graph showing the height of the water level in the pool over time. Be sure to include all of the activities Sylvia did to prepare the pool for the party. Remember that only one activity happened at a time. Think carefully about how each section of your graph will look, labeling where each activity occurs.

3. Does your graph represent a function? Why or why not? Would all graphs created for this situation represent a function?

SOURCE: https://access.openupresources.org/curricula/our-hs-math/integrated/math-1/unit-3/lesson-1/index.html Used with permission from Open Up Resources.

What math ideas did you encounter when you were working on the task? What academic language did you use to describe those ideas?

According to the curriculum guide, the goals for this task include graphing a function to model a situation and interpreting the key features of the graph. Open Up Resources also notes the notation, vocabulary, and conventions relevant to this lesson (see Figure 2.2).

Adding Notation, Vocabulary, and Conventions

Key features of functions:

- *x-intercept – when* $f(x) = 0$
- *y-intercept –* $f(0)$
- *Maximum – greatest y-value*
- *Minimum – lowest y-value*
- *Interval of increase – as x values increase, y-values increase*
- *Interval of decrease – as x values increase, y-values decrease*
- *Rate of change –* $\frac{\text{change in } y}{\text{change in } x}$
- *Constant rate of change is the slope for linear functions*
- *Rate of change is zero when there is no change in y-values, the graph is horizontal*
- *Domain – the set of possible input values for a relationship*
- *Range – the set of possible output values for a relationship*
- *Continuous, discrete, discontinuous*

SOURCE: Reprinted with permission from Open Up Resources (2021).

■ ■ ■ Try This

What language might we expect students to use as they work on this task? Jot down some ideas to add to the list of your own academic language that you used. If you have an opportunity to try the task with students, you might want to transcribe or audio record the language they use. ■

Here's some language we might expect to hear when students are describing their graph while sharing the story they created about the situation:

- "Nothing goes up or down when the pool is being cleaned."

- "Time goes this way (gestures left to right)."

- "A bucket would be a different kind of drop than draining with a hose. It's a dip (gestures straight down) not a drip (gestures diagonally)."

- "Do Sylvia's friends take the buckets out the same time she does, or do they go 1 2 3?"

- "Would we draw the line the same for the fill and the drain? Well, not the *same*, like the same but opposite."

- "The pool needs to be empty before we clean it."

- "Nothing changes during the break—do we just stop drawing and pick up over here?"

- "It can't just go up and up forever though, like at some point the pool is full and you've got water just spilling into the yard."

- "Look like over here at the start (gestures to *y*-intercept) you've got the pool more full than it is at the end. Shouldn't we stop where we started?"

The highly contextual nature of the task makes it likely that much of the academic language students use will draw on aspects of the context, such as buckets, filling, and draining. And when students are huddled in a small group around a graph or guiding a classmate as they're describing the story they created, the academic language they use is likely to include gestures—a hand motion, a part of the graph they're pointing at, or a sweep of the pen. Using context, gestures, and the graph is an example of using and connecting representations, one of the eight National Council of Teachers of Mathematics (NCTM) Effective Mathematics Teaching Practices.

SOURCE: istock.com/Alina Vovk

Compare this language with the vocabulary terms in Figure 2.2. We don't see many of those specific terms in students' language. But they are making use of the ideas using contextual language. We've added annotations using bold type in the bulleted list that follows to those quotes to match the vocabulary ideas to students' academic language.

- "Nothing goes up or down when the pool is being cleaned." **rate of change, interval of increase/decrease**

- "We're draining here and we're filling here." **Rate of change, interval of increase/decrease**

- "Time goes this way (gestures left to right)" **domain**

- "A bucket would be a different kind of drop than draining with a hose. It's a dip (gestures straight down) not a drip (gestures diagonally)." **Rate of change, interval of increase/decrease, continuous vs. discontinuous**

- "Do Sylvia's friends take the buckets out the same time she does, or do they go 1 2 3?" **continuous vs. discontinuous, discrete function, rate of change**

- "Would we draw the line the same for the fill and the drain? Well, not the *same*, like the same but opposite." **Rate of change, interval of increase/decrease**

- "The pool needs to be empty before we clean it." **Maximum/minimum**

- "Nothing changes during the break—do we just stop drawing and pick up over here?" **continuous vs. discontinuous, domain**

- "It can't just go up and up forever though, like at some point the pool is full and you've got water just spilling into the yard." **Maximum/minimum, range**

- "Look like over here at the start (gestures to y-intercept) you've got the pool more full than it is at the end. Shouldn't we stop where we started?" **y-intercept, maximum/minimum**

Going back to Figure 2.2, it's often very tempting to define all the terms students might need before they start an exploration. And there are times when knowing a definition in advance can be really important—for example, if we are starting a lesson focused on properties of trapezoids, the definition of what counts (or doesn't count) as a trapezoid is vital to that lesson's success! But when we define those terms and ask students to start using them, we take away opportunities for students to make use of the assets they might bring in the form of contextual, informal academic language to describe math ideas.

An alternative to defining terms first is to define terms at the close of an exploration. One way to close a lesson using Getting Ready for a Pool Party would be to take the vocabulary terms in Figure 2.2 and to have students annotate their story and graph with as many of the terms as they can. In this way, students bring their assets to the main task in making sense of the context and can connect that understanding with the specific vocabulary and mathematics terminology that will be useful beyond this specific context. For example, a future task might feature a speed versus time graph of a bike ride and we might expect students to start to use terms like *rate of change* and *interval of increase or decrease* in the context of the new task.

STRENGTHENING MATHEMATICS DISCOURSE AMID COMPLEXITY

Consider the ideas in Table 2.2 related to geometry and measurement and the grade band in which they're likely to be introduced.

TABLE 2.2   Geometry and Measurement Terminology Across the Grades

GEOMETRY AND MEASUREMENT IDEA	GRADE INTRODUCED
Triangle	Early elementary
Polygon	Early elementary
Angle	Late elementary
Angle measure	Late elementary
Interior angle	Middle grades
Exterior angle	Middle grades
Sum of the measures of the angles in a triangle	Middle grades
Alternate interior angle	Middle grades
Angle and direction of a rotated preimage of a polygon	Middle grades
Sum the squares of the legs of a right triangle	Middle grades
Arc length of a sector of a circle	High school
Corresponding sides of congruent triangles	High school
Center of dilation	High school

What did you notice about the terms? As we describe more complex geometry and measurement ideas, our names for those ideas become longer and increasingly grammatically complex. For example, "sum of the measures of the angles in a triangle" is a single idea with four different levels of nesting inherent in the phrase (there is a triangle, it has three angles, those angles have measurements, and we're interested in the total of those). As students move into middle and high school, the terminology we use to describe math ideas changes and increasingly features dense noun phrases (Herbel-Eisenmann et al., 2017) and combinations of multiple vocabulary words and terms to generate new meaning.

> **Tip**
>
> Introduce new academic language using multiple modalities—written, verbal, visual, symbolic—to increase access for students.

Navigating these dense phrases can be challenging for all students, especially multilanguage learners. What can we do as teachers to support student understanding and build on student assets? One important idea is using multiple modalities regularly in classroom practice. If students only hear the phrase, "angle and direction of a rotated preimage of a polygon," students may not fully parse the verbal utterance or understand that those ten words refer to a single mathematical idea. Actions such as writing the phrase down, pairing it with a visual representation of the idea, and connecting the phrase to contextual situations when possible all can support students in understanding new ideas and broadens our work from presenting vocabulary to introducing academic language.

Returning to the ideas in the previous section, it's also helpful to understand that different activity structures in the classroom promote different types of academic language. Herbel-Eisenmann and colleagues (2017) introduced the idea of the *Language Spectrum* (Table 2.3) in their work on supporting meaningful mathematics discourse in secondary classrooms.

TABLE 2.3   Language Spectrum

CONTEXT 1: SPOKEN BY A SMALL GROUP OF STUDENTS WITH ACCOMPANYING ACTION OR GESTURE	CONTEXT 2: SPOKEN BY A STUDENT ABOUT THE ACTION, AFTER THE EVENT	CONTEXT 3: RESPONSE WRITTEN BY A STUDENT	CONTEXT 4: WRITTEN DESCRIPTION FROM A MATHEMATICS TEXTBOOK
Student 1: OK, so I think you just take this away from this, and then you just have, like, something on the top. Like, here and here [*points at examples*], there isn't anything left. They all just cancel out. I think that's why the rule works. You can cross out the numbers under here [*points to the denominator*]. Student 2: Couldn't you have, like, more on the bottom?	Student 3: Remember when we had that assignment where we had to write out what all the exponents meant, like three to the fifth power was three times itself five times? And when we did that with the division problems you could cancel out the same amount on the top and bottom? Like, if there are five on top and three on the bottom, you can cancel three of them and just have two left. But we just did that problem with b to the m on top and b to the n on bottom. So, just like we said five minus three is two, you do m minus n and that's what you have left. That's what we got.	When you divide exponents with the same base, like $\dfrac{b^m}{b^n}$, there are m copies of b in the numerator and n copies of b in the denominator. You can simplify this expression because copies of b in the numerator will cancel with copies of b in the denominator. Since $$\frac{1}{b^m} = b^{-m},$$ $$\frac{b^m}{b^n} = b^m \times b^{-n}.$$ When you multiply exponents with the same base, we add the exponents, so $$\frac{b^m}{b^n} = b^{m-n}.$$	In the case of division where the bases of the exponential expressions that are divided are the same, such as $\dfrac{b^m}{b^n}$ where b, m, and n are rational numbers, the result is b^{m-n}. This is a consequence of the multiplication rule for exponents with like bases. $$\frac{b^m}{b^n} = b^m \times b^{-n} = b^{m-n}.$$

SOURCE: Adapted from Herbel-Eisenmann et al. (2017).[1]

----

[1] For more information about these materials, please email the lead author of Mathematics Discourse in Secondary Classrooms (MDISC) at bhe@msu.edu. We acknowledge that the information related to this table is based on the idea of a "mode continuum" from systemic functional linguistics (Gibbons, 2009) and the naming as the Language Spectrum is attributed to the teacher researchers involved in Herbel-Eisenmann's National Science Foundation (NSF) CAREER grant (#0347906). Any opinions, findings, and conclusions or recommendations expressed in this material are those of the authors and do not necessarily reflect the views of the NSF.

(Facilitating meaningful mathematics discourse is one of NCTM's Effective Mathematics Teaching Practices.) There are two important things to note from the Language Spectrum that relate to an asset-based approach to language. First, the pronouns students use, the extent to which nonverbal and nonwritten facets of discourse are present, and the precision of the academic language used change across the communication contexts listed at the top. When students are working in groups, the language is likely to be more contextual and based in shared referents. When students are presenting their ideas to the class or writing an idea up for homework, the tense shifts to third person and more terminology creeps in.

Sometimes we introduce a new piece of vocabulary or terminology and expect to hear students using it right away in their discussions and in their written work. The Language Spectrum helps us understand that the communication context we ask students to engage in can shift the use of those vocabulary terms. Moreover, even if students have used a term one day, perhaps in their written work like context 3, subsequent work that leverages contexts 1 or 2 might shift language back toward more contextual ideas. The goal is not to move to the right on the Language Spectrum; rather, the framework helps us understand and anticipate shifts in language depending on how we ask students to engage.

USING LANGUAGE IN WAYS THAT LEVERAGE ASSETS

Now that we have explored the impact of introducing vocabulary and terminology, as well as the challenges around complexity of language, how might we take these factors into account in our classroom? This section shares some strategies for supporting students in leveraging their language assets in three categories. These suggestions work most effectively when implementing tasks that promote reasoning and problem solving as compared with lessons that center on more procedural practice.

CREATE SPACE IN LESSONS FOR STUDENTS TO BRING THEIR MATHEMATICAL ASSETS

Tip

When launching a task at the start of a lesson, provide opportunities for students to share their initial mathematical ideas before they get started.

When planning a lesson, provide clear and explicit opportunities for students to bring their mathematical assets. Rather than assuming a "blank slate" and starting a lesson with sharing ideas that you hope students will use, instead acknowledge the fact that students come in with ideas about mathematics that are helpful for everyone in the class to hear. At the start of a lesson, it's helpful to create space for students to share those ideas. This can provide a different means for activating prior knowledge than a procedural set of warm-up problems.

> **Tip**
>
> Introduce vocabulary and terminology *after* students have had a chance to make their own mathematical meaning of the content.

We suggest delaying the introduction of specific vocabulary and terminology until later in lessons when students have had opportunities to make meaning and reason in rich problem-solving situations. This approach allows students to use informal language and conceptions—assets that they bring to our classroom—as a part of their mathematical reasoning and avoids the role that terminology can sometimes play as a barrier to understanding. It also gives students opportunities to revise and refine their thinking and explanations as they go, communicating the idea that mathematics is a subject where we develop and change ideas similar to draft writing in an English course (Jansen, 2020).

PLAN FOR SHIFTS IN LANGUAGE WITHIN AND ACROSS LESSONS

As we saw in the Language Spectrum, the nature of the language students use shifts as the context in which we ask them to use that language shifts. Not every situation should sound like the formal textbook-like language in context 4 of the Language Spectrum. When we're planning lessons, we should align our interactions with students and questions for them with the contexts that they're working in. For example, when students are working in small groups, it's natural for the

> **Tip**
>
> As students move from small-group discussions to a whole-class discussion, introduce more formal language *alongside*, not in place of, informal language.

language we use to ask questions and understand their thinking to be more informal and contextually based: "What's going on right here in the graph? Why did you add these numbers together (pointing to student writing on a page)?" When writing lesson plans, sometimes it's tempting to use more formal mathematical language; questions while students are working in small groups might be better served by less formal language that meets students where they are.

Planning for shifts in language across a lesson is also important. The language we use to ask questions of students in small groups is likely to be different from the language we use in a whole-class discussion. When we're bringing mathematical ideas together and connecting them, this

> **Tip**
>
> When you plan your lesson flow, write down the specific moments (like small-group work) when you want to use and promote informal language and specific moments (like a whole-class discussion) to introduce formal language.

may be a time to move toward more formal language and to introduce specific vocabulary and terminology. We should plan explicitly for changes in how we ask questions and interact with students depending on whether they are working individually, in small groups, discussing as a whole class, and what our goals are for each of those segments of a lesson.

BROADEN WHAT COUNTS AS ACADEMIC LANGUAGE

Tip

Use gestures, inscriptions, and visuals as a part of academic language for you and your students.

Tip

As inspired by Peter Liljedahl's *Building Thinking Classrooms* (2021), use nonpermanent vertical surfaces on walls to make it easier for students to share their work when they are explaining, writing, and gesturing.

Tip

"De-front" your classroom by positioning student desks facing one another in pairs or small groups (Liljedahl, 2021).

Tip

Encourage students to talk to one another (rather than talking to or through you) during whole-class discussions by standing to the side or in the back of the classroom.

What did you think of when we first used the phrase *academic language* in this chapter? What are you thinking about when you see that phrase now? Hopefully as a result of the ideas in this chapter, you've broadened your idea of what counts as academic language. Academic language isn't just the formal vocabulary and terminology of math. It includes informal language that students bring from their own experiences to describe math ideas. It includes contextual language that we all use when we're working on real-world problems.

An important piece of leveraging student language assets is to make space for an expanded and inclusive concept of academic language. We should make space in our lessons for gestures, inscriptions, and visuals. This includes making physical space, including centering students when they explain so gestures and visuals are accessible to all, and creating board and/or projector space both for students to use in sharing their ideas and for us as teachers to record important ideas (Liljedahl, 2021).

Relatedly, as teachers, we need to consider the power and privilege that comes with our status in the classroom. The ideas we choose to highlight are consequential to students because of our role. We need to think about how we explicitly mark and value informal math language and an expansive concept of academic language in the moment-to-moment work of teaching. Planning in advance helps us do this well, but we also need to think about how this unfolds in the moment.

PUTTING IT ALL TOGETHER: REVISITING THE POOL PARTY

Let's pull together the ideas we've discussed in the chapter and think about how we'd plan for the Getting Ready for a Pool Party task with academic language in mind. As you work on the next activity, put the population of multilanguage learners that your school or classroom serves at the center of your discussion. In addition to the asset-focused academic language practices you're integrating into your planning, how are you specifically supporting this learner group?

■ ■ ■ Try This

Plan an imagined lesson using the Getting Ready for a Pool Party task. As you plan, discuss the specific actions you will take in different phases of the lesson (launch, small-group work, whole-class discussion) to leverage student assets related to language.

Consider specific ways to support the multilanguage learners that your school or district serves.

After you've discussed your plan, read on to learn how we thought about the task. ■

LAUNCHING THE TASK

In planning for the lesson, I (Mike) launched the task by asking students to think about draining, cleaning, and refilling a pool. I asked students to think about those situations and where they might make use of mathematical ideas to get that work done. I was mindful of the vocabulary list that's provided and made sure to revoice important ideas students shared that relate to those terms. I wrote down some key phrases or ideas on the whiteboard so that we could return to them later, and they were available for students to reference in small groups. Then I provided 3–5 minutes for students to individually get started on the task and then transitioned to small-group work. The individual work time wasn't enough to complete the task but afforded students opportunities to use some of their assets before bringing those ideas to a small group in which they would change and evolve.

SMALL-GROUP WORK

I provided about 20 minutes for small groups to grapple with the task. As I walked around the class to listen to the discussion, I made a list of the informal and contextual language that relates to those key vocabulary ideas. I listened to groups and asked clarifying questions, making use of their instructions and being attentive to gestures. I also made sure that all students in the group had a chance to use their language and tuned in specifically to my students with special language needs like emerging multilingual students or those who may be more comfortable writing than talking. I made notes about which groups are talking about the key aspects of academic language that I wanted to bring to the whole class and used those notes to organize my whole-class discussion to bring those ideas forward.

WHOLE-CLASS DISCUSSION

After I monitored the groups and had a good sense of which groups could help share their thinking about specific math ideas embedded in the task, I started the whole-class discussion. Given the importance of the context and the graph in this task, I made space for students to both speak about their solution and display aspects of it either with a document camera or with other digital technology.

During the discussion, I listened for the sorts of words and phrases that we noted in our earlier discussion. When a student used a phrase that represented a key math idea, like "goes up," I asked if anyone talked about or showed that same idea in a different way. I made space for gestures and visuals along the way and jotted down key phrases to add to our list from the start of the lesson.

To close the lesson, I provided the list of terms shown in Figure 2.2 and asked students to do two things based on their work on the task. First, they needed to annotate their work with places that reflected each term. Then for one term that their work didn't reflect, write one to two sentences about how they could use that idea in the context of the task.

LESSON SUMMARY

To summarize, students bring a wide range of mathematical assets to the classroom. So far in this chapter, we've talked about different ways to identify those assets and specific strategies we can use in the planning and teaching of a lesson to use those assets. Although many of our decisions in the moment of teaching can help leverage student assets, thoughtful planning helps us to intentionally create space and time within a lesson for leveraging students' language assets. In the last section of the chapter, let's talk about how we can extend that planning to attend to students with specific language needs.

Digging Deeper
More on Multilingual Learners and Universal Design for Learning

Emerging multilanguage learners have special language needs in the math classroom. The ideas we've discussed already in this chapter, such as delaying the introduction of vocabulary and terminology and a broad conception of academic language, are important for all learners, but particularly important for multilanguage learners. It's important to make sure we recognize the assets that multilanguage learners bring and provide them with opportunities to use those assets, even if as a teacher we are not fluent in their native language.

Rachel Lambert's Universal Design for Learning (UDL) Math (Lambert, 2021, 2024) framework introduced in Chapter 1 is particularly useful as a way to think about leveraging language assets for multilanguage learners in the secondary math classroom. Let's talk about one facet of the framework in each of the three categories—Engagement, Representation, and Strategic Action. ■

ENGAGEMENT: MEANINGFUL MATHEMATICS

Earlier in the chapter, we discussed planning lessons in ways that started with opportunities for students to bring their language assets and informal math understandings to the table. In the Pool Party task, this entailed providing students opportunities to discuss where they saw mathematical ideas in preparing a pool for a party. For our multilanguage learners, recruiting interest in ways such as this means being thoughtful about access. How might we plan to support multilanguage learners in understanding and being able to translate the key aspects of a context? What contexts are likely to be familiar to our students based on their cultural backgrounds, and which may require additional work to support their understanding? In the example of the Pool Party task, learners coming from cultures where home-based pools are not common are likely to be disadvantaged in talking about the math involved in draining, cleaning, and filling a pool. In these situations, video clips of the activity or careful prework in translating the key activities (like cleaning, draining, and filling) can support access. And as with many aspects of UDL, providing support for particular students allows for good access strategies for all students!

REPRESENTATION: INVEST TIME IN CORE IDEAS AND MULTIMODALITY

The UDL Math framework notes the importance of investing time in core ideas. Our strategy of delaying the introduction of vocabulary until after mathematical meaning-making keeps the focus on the core ideas before the introduction of vocabulary and terminology. Another aspect of the Representation category is multimodality, a strategy we highlighted in the section but that has particular importance for multilingual learners. The inclusive conception of academic language that includes gestures, inscriptions, and visuals supports making language and symbols accessible to multilanguage learners.

STRATEGIC ACTION: UNDERSTANDING SELF AS A MATH LEARNER

This facet of the UDL Math framework notes that we should support opportunities for students to reflect on their own development as mathematics learners (Lambert, 2024). With respect to students' language assets, this is when the idea of the Language Spectrum is particularly helpful. The different contexts for discussing mathematics, from informal conversations among students about a task to reporting to the whole class to a formal written explanation, can provide opportunities for students to learn about themselves as math learners and for us as teachers to support their strategic development. As we noted in discussing the Language Spectrum, the goal is not to move from left to right but to understand how language shifts and changes as we move in both directions on the Language Spectrum. Multilanguage learners are faced with the challenge of developing their language skills simultaneously with their content knowledge. Providing students with multiple opportunities to move on the language spectrum and to reflect on their language use while making sense of a new mathematical idea can be particularly helpful and supportive for multilanguage learners.

Reflect, Apply, Transform

To conclude our work in this chapter, let's think about our beliefs and how they relate to opportunities to position student language on the asset-based side of the deficit-to-asset continuum. When I first started teaching mathematics, I replicated the way I had been taught, which included starting off with key definitions like the notes shown in Figure 2.1. It took a very long time for me to stop and deeply consider *why* I had adopted those language practices and how that was both affording and restricting access to certain students in my class. When it came down to it, my beliefs about what was necessary for students to begin their mathematical learning were surfaced, and many of those beliefs were unproductive and not reflective of how students learn. In that spirit, the Reflect, Apply, Transform activity below provides you with some writing and discussion prompts to jump-start that reflective process. If you're working by yourself, you might use these as writing prompts. If you're working with a team, we encourage you to write some initial thoughts and then have a discussion with your colleagues about what you wrote.

1. What do you consider to be the most important things for students to be able to do with respect to academic language? Why are these important to you?

2. When a student uses informal mathematical language, what's your initial reaction? What reasons might students use informal language rather than formal language, even if they've already learned the mathematical term?

3. Think about the contexts you've used in recent problem-solving lessons. To whom are those contexts likely to be familiar? To whom might those contexts be less familiar?

4. Reflect on a recent lesson that you taught. When were there opportunities for students to leverage the assets that they brought to the classroom that day? When were there opportunities to listen to and build on the assets of others?

5. How might the use of students' language assets be explicitly valued in the work we do assessing student performance?

CHAPTER 3

Empowering Students Through Language

In this chapter, we'll build on the work we did in Chapter 2 thinking about asset-based language. We'll look at how our classroom norms focused on language can frame students' opportunities to use asset-based language. We'll develop some ways to think about how a productive set of classroom norms can effectively mediate language use. At the end of this chapter, you should have a new set of tools and perspectives to help you consider how your norms can foster asset-based learning environments. Because norms can vary from one teacher's classroom to the next, this chapter is particularly good to use with a collaborative group such as your professional learning community (PLC).

> *"The norms and cultures of the classroom . . . are highly necessary elements in establishing meaningful interactions that support mathematical discourse."* *(Bennett, 2014, p. 20)*

Questions to Consider

1. How do your classroom norms support asset-based or deficit-based language?

2. How can your classroom norms be revised to afford access to a wider range of students and promote their asset-based language?

3. What implicit norms can be made explicit by posting or discussing with your class?

ALIGNMENT EXERCISE: ANALYZING CLASSROOM NORMS

Mx. Cooper's PLC team was getting ready to start the new school year. The team had noticed that some students had struggled with differences in norms and standards from one classroom to the next at the end of last year, so all the team's teachers were bringing their classroom norms to a summer PLC meeting with the goal of looking across the set and thinking about revisions. The team didn't want everyone's norms to necessarily be exactly the same, but they were hoping to better understand where there were areas of resonance and where there might be conflicts.

■■■ Try This

Mx. Cooper's norms are shown in Figure 3.1. Read through their norms. For each one on the list:

- record how you think the norm could support the development of asset-based language use in the classroom and

- record how you think the norm could promote a deficit-based perspective on language in the classroom.

FIGURE 3.1 Mx. Cooper's Classroom Norms

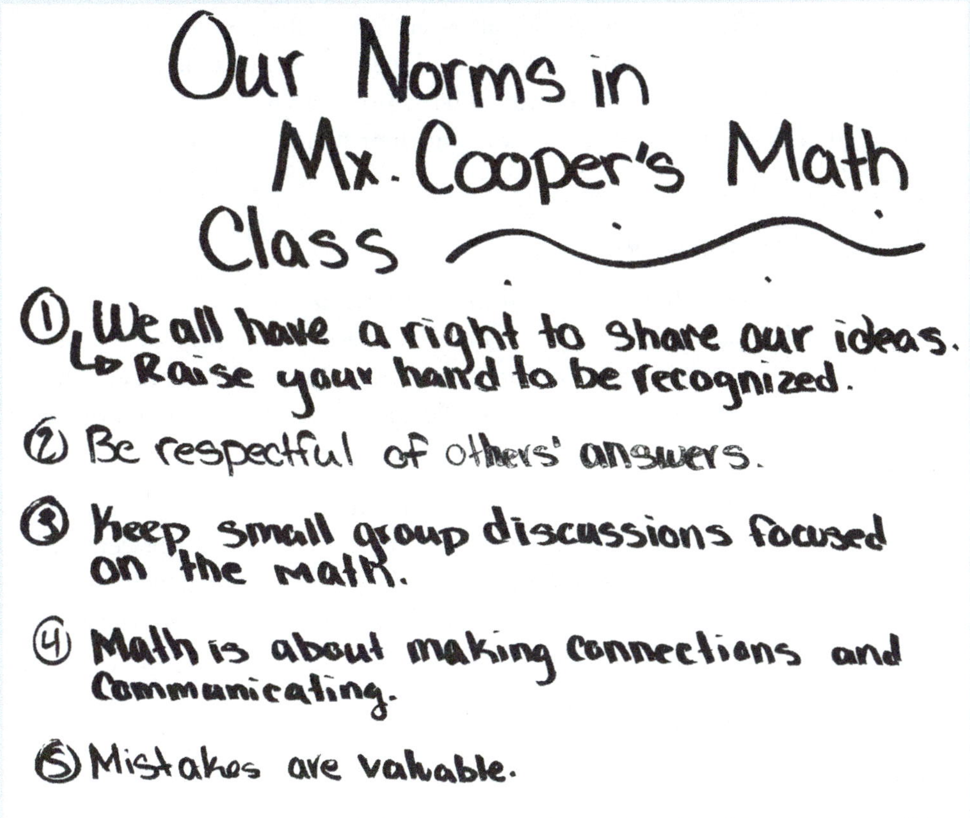

Our Norms in
    Mx. Cooper's Math
    Class

① We all have a right to share our ideas.
    ↳ Raise your hand to be recognized.

② Be respectful of others' answers.

③ Keep small group discussions focused on the math.

④ Math is about making connections and communicating.

⑤ Mistakes are valuable.

What did you notice about each of the five norms? All five felt familiar. I (Mike) probably have used versions of some of these myself, and on the surface they may seem very positive. But let's think about how our students might make sense of these norms and where they might reflect asset-based or deficit-based perspectives on language.

1. ***We all have a right to share our ideas. Raise your hand to be recognized.***

   From an asset-based perspective, this norm communicates to students that their ideas are valued. At a broad level, their language and thinking is welcomed in the classroom. In that sense, this norm sends a great message.

   The second sentence of the norm is certainly well intentioned: Hand raising is one way we've used to bring order to what can be the chaos of classroom conversations, particularly as students are still learning how to interact appropriately with one another. But who might hand-raising exclude? Does this norm limit students' abilities to engage in spontaneous conversations with one another during a whole-class discussion? In some cultures, overlapping speech and adding-on is viewed as a collaborative and welcome activity (Tannen, 2021); other cultures adhere more to a don't-speak-unless-spoken-to set of practices. Does this norm send a specific message to students from cultures in which overlapping speech and dialogue is valued? Or for whom raising their hands could be anxiety provoking?

2. ***Be respectful of others' answers.***

   This norm is one that we've seen in many classrooms, and it's one with which few would disagree. There is an asset-based flavor to this norm: By emphasizing that others' answers are important and worth hearing, students are invited to consider perspectives that may not be their own.

   What aspects of this norm could be interpreted in a deficit-based way? Does the focus on others' *answers* perhaps implicitly signal that the answer is what is valued in class, not the thinking and reasoning? It may be worth considering whether a revision of this norm could explicitly embrace a wider range of students' mathematical asset.

3. ***Keep small-group discussions focused on the math.***

   I'm sure we all understand *exactly* what Mx. Cooper had hoped to accomplish with this norm. Particularly in middle and high school, small-group discussions can get off topic quickly! You could look at this norm as encouraging students' mathematical assets to come forward.

Let's think about what sort of discourse this might sideline. Our hopes and dreams for our math students are that they'll develop tools in math class that help them make sense of and mathematize the world. We recognize the importance of real-world context in the tasks that we give students. So consider: Does a norm like this restrict students' abilities to bring real-world assets into their conversations about mathematics?

4. *Math is about making connections and communicating.*

Here's another norm that seems like it's strongly asset focused on the surface, and there is a lot here that invites students to bring their mathematical assets to their work. This norm expresses value for a wide range of contributions and helps students understand that math is about more than getting the answer.

That statement may remind you of norm 2. How does this norm relate to that one? Also, we want to consider the wide range of students in our classrooms. How does this norm invite students who may be more deliberative as compared with immediately contributing to a conversation? Could this norm be implemented in ways that privilege faster thinkers and faster talkers? The norm itself may not explicitly express these values, but it's not hard to imagine how students might interpret it or a teacher might implement it.

5. *Mistakes are valuable.*

The value of mistakes, and more broadly productive struggle, has been well documented in recent writing about math teaching (e.g., SanGiovanni et al., 2020). This norm positions mistakes, which we all make, as an asset in the learning of mathematics.

Similar to some of the other norms we've examined, this norm doesn't have an explicitly deficit-based flavor to it. But we can wonder how students might make sense of this. What specifically should a student do with a mistake they make? How might we support students in coming to see those mistakes as valuable?

This last question is perhaps the most critical. Our educational community often discusses the idea that mistakes are valuable, but sometimes this contrasts with a persisting focus on correct answers. It's important for us to think about how to back up statements like *mistakes are valuable* with clear, visible, and actionable practices in our classrooms. For example, we might use Jansen's (2020) notion of rough draft math to position early thinking that may contain inaccuracies or imprecisions as valuable. Saying that learning is a process and the evolution and revision of our thinking is important is a good first step, but we need to back this up with tangible actions that demonstrate to students that we truly value the evolution of their thinking.

We hope this exercise demonstrated a few things about classroom norms and their relationships to language. First, norms themselves aren't generally asset or deficit based, but the ways in which teachers and students might interpret and enact those norms can be! This point is a critically important one as we often focus a great deal of attention on creating norms (individually or collaboratively with students) but perhaps less time reflecting on their implementation. How often do we check in about how our students are making sense of the norms?

Second, considering the norms from multiple perspectives is important. Different norms may mean different things to different students. How are the norms empowering some of our students and perhaps marginalizing others? This also relates to the third point: Who is given a voice in creating the norms? When we work collaboratively to create norms at the start of the year, who feels empowered to contribute and who may not? How do these norms get revisited and adjusted during the course of the year?

■ ■ ■ Try This

Before moving on in the chapter, we'd like to invite you to analyze your own classroom norms. You can do this individually or, like Mx. Cooper, in collaboration with your PLC or another group of colleagues. For each norm, you might consider the following questions:

- How does this norm reflect asset-based or deficit-based perspectives?

- How might my students make sense of and implement this norm?

- How do I implement and support students in meeting this norm?

- What aspects of this norm might invite refinement or revision? ■

HOW DO CLASSROOM NORMS REFLECT ASSET-BASED LEARNING ENVIRONMENTS?

The last chapter took a close look at the range of academic language used in math class and how that language relates to the deficit-to-asset continuum. This chapter broadens our lens on language to consider how we support students in developing and using language in asset-based ways. The classroom norms that we create, publicly post, and refer to throughout the year frame the language students choose to use in math class, what is considered appropriate for math class, and what ideas are not welcomed.

CLASSROOM NORMS AS MEDIATORS OF LANGUAGE USE

As we saw in the Alignment exercise, norms themselves may not be asset or deficit based, but the implementation of those norms by teachers and students can amplify or dampen the assets students bring to math class. Inviting students to bring their assets doesn't mean that anything goes, so we don't want you to take from this chapter that fewer norms are better. We need boundaries and guardrails in our classrooms to make sure all students feel it is a safe and productive space for learning and so that we as teachers can move our class collectively toward learning goals. So the work of setting and revising our norms has to consider the tension between what is invited and what is discouraged and how that maps onto the assets we want students to bring into the classroom. It's an imperfect and messy process, and to help us think about it, we're going to use some illustrative cases of norms and the different kinds of asset-based language we want students to bring to their learning.

> **Tip**
>
> Post classroom norms developed with students that are empowering and encourage them to share their assets.

CASE 1: MR. CAI'S CLASSROOM

Mr. Cai had been encouraging his students to share their ideas in his seventh-grade math class with some success. Students were volunteering, contributing, and sharing their thinking and reasoning. Mr. Cai had noticed, however, that students were rarely talking to one another about their ideas. Even in small groups, students would raise their hands to get Mr. Cai's attention and ask a question or make a comment about another student's thinking that would be better directed to the student. In whole-class discussions, Mr. Cai noticed that even when students would comment on another student's solution that they were discussing, the students would direct the comment to Mr. Cai. Mr. Cai was often having to say things like Xander, what do you think about Shonda's comment to your solution?

After a few weeks of redirecting students back to one another, Mr. Cai took a look at the classroom norms he had posted. Two norms caught his eye, and he wondered how they might be influencing students and their willingness to engage in student-to-student discourse.

Classroom Guidelines
for Mr. Cai's Math Class:

· Take turns speaking and make sure everyone has an opportunity to contribute to small groups

· Critique the idea, not the person.

■■■ Try This

Pause here and consider . . . how might these two norms serve to promote and restrict asset-based language in the form of **student-to-student talk**? Jot down a few of your thoughts; if you're working on this chapter with a group, take a few moments to discuss the case and what you notice. ■

Mr. Cai thought carefully about these two norms. He had established them because he wanted to make sure all students had a voice in conversations and that students' emerging mathematical identities were considered so that they didn't feel critiqued. But in thinking about why students might not be talking with each other, he identified a few possibilities. For the first norm, he wondered whether the phrase *Take turns speaking* implied for students that someone should be governing who takes the turns and how. When we talk about taking turns, we might think about games that have rules and specific ways in which turn taking happens. Were students looking to him to govern whose turn it was to speak, particularly if multiple people might want to speak at a time? This may have inadvertently shut down student speech, at least for some students.

In looking carefully at the second norm (critique the idea, not the person), Mr. Cai continued to think this was very important. But he remembered that when students were offering those critiques of a student's solution, they were directing them *to him*. Was this the way students were interpreting this norm? If students didn't direct critiques to another student, was that how they were thinking about not critiquing people?

Mr. Cai made a decision to modify the first norm to be, "We should all have the opportunity to offer our thoughts about math to one another and contribute to discussions." He decided to leave the second norm as it was written but planned to take some time before their next discussion to discuss what the norm meant and to do some explicit modeling of *how* we can critique one another's ideas by speaking directly to one another.

> **Tip**
>
> Plan explicitly to revise your classroom norms throughout the year.

CASE 2: MS. JENKINS'S CLASSROOM

Ms. Jenkins's high school Algebra I class of ninth graders started the year really hungry to know whether their answers were *right*. When she invited them to share their thinking, or even just to give a show of hands of whether they agreed with an answer, there was hesitation almost across the board. Students began pulling her aside as she was asking these questions pointing at their paper and whispering, *but is this right?* She had consistently emphasized that their thinking was important and that mistakes were a part of the learning process, but getting this class to that point was a work in progress and, at times, a struggle. The students seemed really fixated on the answer being the most important thing. When she asked them to share their reasoning, they'd often just say the answer they got or if they shared more than that, they would just share the mathematical steps that they did but not *why* they chose the steps that they chose.

Ms. Jenkins looked at her classroom norms after one particularly challenging Friday afternoon working on linear functions to try and determine what she could do differently to emphasize that reasoning was important.

Original Algebral Norms

✿ Show your work. It's important for us all to understand the mathematics that you did.

✿ We all make mistakes and they help grow your brain.

✿ MATH is about communicating

✿ Your reasoning is as important as your answer.

■ ■ ■ **Try This**

Pause here and consider . . . how might these norms serve to promote and restrict asset-based language in the context of students **sharing their mathematical reasoning**? Jot down a few of your thoughts; if you're working on this chapter with others, take a few moments to discuss the case and what you notice. ■

Ms. Jenkins noted a few things in the list of norms for her Algebra I class that they had developed collaboratively. She liked everything there and felt that all four of those norms could be looked at as welcoming students' language assets into the classroom. They embodied the idea that mistakes are valuable, that showing our work is important, and that communicating and reasoning are valued. But she tried to press herself to think about how those norms might be restricting students in communicating their reasoning.

She took out some recent student work and looked carefully at it with the norms in mind. Much of what she saw were a lot of procedural steps as students were solving and graphing linear equations. She also saw a lot of work in pencil with evidence of erasing. How were students interpreting the idea of showing work? Was it perhaps just the steps? And was it clear to students how their mistakes would be helpful given that there was evidence of them revising them to make the mistakes disappear? Maybe this was a place where there could be some recalibration with the class about what "work" means and how mistakes could be helpful. She also noticed that when she asked students to explain their answers, even with contextual problems, they were just restating the mathematics that was written in symbols with words. Sometimes the explanations reflected what students had done mathematically but didn't in any way seem to help them challenge a piece of computation that had gone astray. Perhaps the norm about reasoning being *as* important as your answer needed some revision?

After reflection and some discussion with the class, Ms. Jenkins came to a revised set of norms:

Revised Algebra 1 Norms

❀ What you write on the page or share in class should help us understand what you did and why you did it.

❀ When you make a mistake, share how you knew it was a mistake and how you adjust your approach

❀ Sharing your reasoning should help us understand why your answer is reasonable.

CASE 3: MRS. ULION'S CLASSROOM

Mrs. Ulion's eighth-grade class spent a lot of time making sense of rich problems in small- and whole-group settings. She wanted her classroom norms to reflect the idea that all students in a group, and all of us in the whole class, were accountable to one another to ensure that everyone understood and could explain the mathematics being studied. She had created her norms to try to reflect an asset-based approach to student language:

GUIDELINES
to
effective collaboration
in Mrs. Ulion's class:

1) Make sure everyone in the group can understand & explain groups thinking.

2) All students come to whole group discussion with something to contribute from small group

These norms had been working fairly well—discussions were boisterous and mathematically robust. But over time, Mrs. Ulion had noticed that some students always had to be encouraged (sometimes gently, sometimes strongly) to share their thinking. She would check their understanding in small groups, and in most cases, they could explain what the group was doing. But sharing in front of the whole group was challenging for them for reasons that weren't clear. Often, other members of the small group would jump in and help out, sometimes explaining that very student's thinking themselves. Mrs. Ulion worried about these students deeply and wondered how she could better attune her norms to support the development of their mathematical identities.

Pause here and consider how these two norms might serve to promote and restrict asset-based language in the service of **students' developing mathematical identities**. Jot down a few of your thoughts; if you're working on this chapter with a group, take a few moments to discuss the case and what you notice. ■

Mrs. Ulion realized that no set of classroom norms could explicitly account for each student, the disposition toward mathematics that they brought to the classroom, and ways to support their identity development as knowers and doers of mathematics. But she also noticed in looking at the two guidelines posted that they were perhaps overly weighted toward the group. Could the norm be revised to better support students who may be more introverted or have social anxiety to share their thinking? Would individuals who were still working to see themselves as knowers and doers of mathematics read the guidelines and see a role for themselves in small- and whole-group discussions?

She rebalanced her norms in the following way:

GUIDELINES
to
effective collaborations
in
MRS. Ulions class
_ _ _ _ _ _ _ _ _ _ _ _

1) Everyone in the group should understand one anothers thinking

2) Acknowledge & credit contributions to the group

3) Give time & space for each person to collect their thinking

4) Contributing to discussions includes sharing solutions & asking great questions

In addition, she instituted some new routines as students were working in small groups. She asked small groups to spend a little bit of time before the whole-group discussion reminding themselves of the contributions everyone had made and rehearsing to one another what they wanted to say in the whole-group discussion to support everyone in being prepared. This took up a little bit more class time to do, but it resulted in stronger contributions when the class moved to a whole-group discussion.

REFLECTING ACROSS THE CASES

The teachers in all three of these cases had a strong set of classroom norms that were intended to elicit students' asset-based language in math class. They had the best interests of their students at heart when creating those norms. But they also went the extra mile in continually monitoring places where the norms were effective in eliciting asset-based language and places that represented challenges. As we saw in each case, the teachers identified specific ways that they wished to empower their students in eliciting their asset-based language and made adjustments to their norms that targeted those areas. Mr. Cai looked to adjust norms to increase student-to-student discourse in his classroom. We see Ms. Jenkins adjusting her norms so that it was clear that asset-based language about the mathematical processes rather than just the answers was valued in her classroom. And Mrs. Ulion rebalanced the focus on group discussions to more explicitly acknowledge and credit individual student contributions, as well as small- and whole-group accountability.

Creating classroom norms (ideally in collaboration with students) is a great way to frame what you'd like your classroom discussions to look and sound like. And just as our students evolve during the course of a year, our norms might need to shift as well. We encourage you to reflect regularly on your classroom norms, revisit and adjust them as needed, and think not just about who the norms are reaching in bringing their asset-based language to the classroom but also about who might be implicitly excluded or discouraged from bringing those assets to the table. Our Digging Deeper section focuses on a set of students who may benefit from specific attention to norms and empowerment.

Digging Deeper
Students Who Need Processing Time

Strengthening student discourse in the math classroom is an important and admirable goal! One thing we often notice is that when we open up our classroom practice to more opportunities for students to bring their language-based assets, some inequities naturally arise. Some students will be eager to start talking and sharing their thinking. We want to honor that willingness and enthusiasm and balance it with the risk that those eager voices might limit the space that other students have to make contributions. As teachers, it's important for us to develop routines to explicitly support and encourage this group of students and to create time and space for their contributions.

Students may be reticent to share for numerous reasons, and it's important to think about those underlying reasons to craft strategies that will be effective. Let's talk about a few groups that benefit from additional processing time and then look at some strategies we can use to level the playing field for these groups. ■

STUDENTS WHO ARE INTROVERTED

Students who are introverted may be less likely to speak up in a small- or whole-group conversation. As a self-identified introvert (Mike) myself, I often enter a discussion situation taking a listening stance first and figuring out where I might best and most productively enter the discussion. Interacting for extended periods of time with groups can be particularly exhausting for introverts; they often need time on their own to recharge their batteries and re-engage. If we think about the pace of a school day that might feature seven or eight different classes and lots of discussions, our students who are more introverted may not always be willing or able to jump into a boisterous conversation, especially at the end of a school day!

> **Tip**
>
> To support students who may be more introverted, ensure private, quiet think time for individuals before moving straight into small-group discussions of a math task. Encourage students to write down initial thoughts they'd like to share during this independent think time, so they are better prepared to transition to group discussion.

STUDENTS WHO ARE NEURODIVERSE[1] AND STUDENTS WITH AUDIO PROCESSING DELAYS

Some students who are neurodiverse can sometimes find dense audio environments challenging. Similarly, students with audio processing delays or hearing impairments may struggle to hear or parse their peers' contributions in a small group, particularly in a classroom in which multiple small groups are conversing at the same time. This may restrict students' access to the conversation as they might not accurately hear other ideas being shared or may shut down due to the inability to keep track of the flow of ideas.

Tip

Consider noise-canceling headphones, a quieter section of the room, or a quick visit to the hallway to provide a break from small-group discussions.

How might we support students who are neurodiverse and students with audio processing challenges? Multiple representations are particularly helpful for these students. We should work to ensure that the tasks we provide have opportunities to use multiple representations and that students have the resources available to them (e.g., graph paper, manipulatives, digital tools) to make use of those representations. We should also focus on encouraging all students to use those representations to provide access to their ideas to others.

In addition, we should think about ways for students to signal that they need a break from a small-group discussion. Providing a mechanism for students to disengage and re-engage from that discussion in short bursts can give students who are struggling to track a conversation a chance to regroup.

STUDENTS WHO ARE QUICK TO RESPOND

But wait . . . I thought this section was going to be about supporting students who may be reticent to engage in a conversation? And it has been . . . but let's take a moment to consider our students who are eager to contribute a first idea to the conversation and think about what they have the opportunity to do in small-group discussions. For students who are quick to offer that first idea that a group takes up and builds on, it's important to recognize that they may miss out on the value and

---

[1] The term *neurodiversity* was popularized in 1998 by sociologist Judy Singer, who identifies as on the autism spectrum but objected to being labeled as having a disability because of it. When we use the phrase *students who are neurodiverse*, we intend to be inclusive of students with variations in brain functionality. This may include students with conditions that may be traditionally labeled as disabilities, such as dyslexia. Not all students who are neurodivergent will identify as having a disability, and not all students with disabilities will identify as neurodiverse. We encourage you as a reader to consider the discussion of students who are neurodiverse as broadly and inclusively as possible.

joy of building on someone else's idea. What might we do to support our traditional early contributors in having this opportunity?

One strategy here is to explicitly call out and discuss the value of both offering new ideas and building on the ideas of others, as both are a part of meaningful mathematics discourse. It's important for students to see that these are different ways of engaging, and being able to do both makes us better mathematical communicators. Periodically, we might encourage students to think about the last time they built on someone else's idea and ask from time to time for students to specifically take that role in a conversation. This recommendation can be made generally to the whole class or specifically to particular students who have frequently made first contributions. And although we might think about the NCTM Effective Teaching Practice of *posing purposeful questions* as largely a teacher move, helping our students see the value of posing purposeful questions to one another can give students who may be quick to offer their mathematical opinion another way to engage with their peers.

BEING WISE AND REFLECTIVE ABOUT OUR TENDENCIES AND BIASES

When I (Mike) started out my secondary math teaching career, I knew that my preservice teacher preparation program had emphasized the diversity of student thinking and paying attention to the variety of ways students might think about a math idea. Even with this focus, I ended up building my classroom norms, language, and instructional routines in ways that centered students who thought and spoke like I did as a student. This tendency is certainly natural, particularly as I was the only eighth-grade mathematics teacher in the school that year and I didn't have easy access to thought partners or other classrooms to observe. It's important that we push ourselves beyond privileging students who think and talk like we do to be more inclusive and better leverage student assets. Listening to students and inviting them in to conversations about classroom norms and mathematical language is a crucial part of doing this well. In addition, colleagues both near and far can help shape your thinking about classroom norms in ways that go beyond your own perspectives and thinking. These actions help us move toward the principles embodied in the Universal Design for Learning (UDL) Math framework (Lambert, 2024) and provide broader access for each and every learner.

Reflect, Apply, Transform

Now that we've looked at several classroom excerpts and thought about the relationships between norms, students' language assets, and specific populations of students, we'd like to invite you to think about your own classroom norms.

If you don't have a set of norms or classroom guidelines posted, think about the norms that you may have implicitly or verbally communicated with your class. Identify two norms from your current list. Analyze the two norms in terms of how they might be implemented by you and your students in asset-based and deficit-based ways and how they might support or inhibit asset-based language.

Once you've done this exercise, revise the norm to afford more access. If you'd like to eliminate the norm entirely, that's fine too, but think about a new norm to replace it.

Finally, identify an asset-based norm related to classroom language that's implicit in your practice—something you value about students' language assets but perhaps have not posted or said out loud. Consider making that norm explicit by posting it or discussing it with your class.

1. How do your classroom norms support asset-based and deficit-based language?

2. How can your classroom norms be revised to afford access to a wider range of students and promote their asset-based language?

3. What implicit norms can be made explicit by posting or discussing with your class?

CHAPTER 4

Unpacking and Evolving Our Language

In this chapter, we'll complete our work on asset-based perspectives on language by considering the ways that we talk about students, student learning, achievement, and progress in math. In this discussion, the school community and the out-of-school community collide—the Western culture has a lot of ways of talking about mathematics that may not align well with an asset-based perspective on language. This chapter extends us beyond our classroom contexts and includes discussions we may have as teachers with one another, with building and district leadership, and with members of our communities. Our overarching goal in this chapter is to examine our language choices and plan for how we can evolve in asset-focused directions.

> *"Deficit narratives and student labeling are central aspects of the historic legacy of underserved students having less access to a challenging education in mathematics than more privileged students (Kitchen et al., 2016; Kitchen & Berk, 2016)." (Kitchen et al., 2017, p. 5)*

Questions to Consider

1. What are aspects of language that you've heard in your building or district, or used yourself, you'd like to change?

2. How do students make sense of the labels used in the school and district with respect to their math identities?

3. In the past, how have you disrupted deficit language that you hear in your school or district?

ALIGNMENT EXERCISE PREAMBLE: SPECIAL EDUCATION

Let's start with a brief story from one of my (Mike's) colleagues and friends, Dr. Kate Johnson. Kate and Mike have worked together on several teacher professional development projects over the years. In one of those projects, Kate was leading a professional development session when the following happened:

"Several years ago, during one professional development session, [Kate] was chatting informally with a group of mathematics teachers as they got to know one another (as they were from several different schools and school districts). The conversation turned toward special education students and the difficulties teachers can face while trying to meet the needs of diverse classes that have high numbers of students with Individualized Educational Plans (IEPs) in them. One teacher stated in a very matter-of-fact way, 'Special education students are the ones that drive you to drink.'

How might you respond?

Would you remember overcoming challenges with a particular student?

Would you, perhaps, laugh it off?

Would you remember a student you taught who made you feel this way?"

Johnson & Fonbuena (2023, p. 145)

The questions posed are important and challenging. Hearing language with which we might not be comfortable, that frames students in particular ways, is both relatable and difficult. How might we note this language publicly? How might we disrupt this language? What does a journey look like from a comment like this to a more asset-based framing of students, and how do we take that journey?

A word of caution before we begin. One thread we'll be discussing throughout this chapter are the words we use to describe groups of students with particular experiences or characteristics. We'll be thoughtful about explaining what words we use and how we're using them. In a few places, we're going to talk about terms we used in the past, why we don't use those anymore, and how they reflect deficit thinking. We'll be careful to explain why we're using the terms we are using when we are using them and for what purposes. If you're working with a group, in some exercises, we'll ask you to talk about how language has changed. It's possible that terms will arise that may be uncomfortable. If you're embarking on the work of this chapter with colleagues, you may wish to preview the content beforehand and set some norms for how to engage in these activities. We'd encourage you to invite people to enter the conversation where they are and to be honest with one another about thoughts, feelings, and reactions to the discussions.

ALIGNMENT EXERCISE: CHANGES IN OUR LANGUAGE

The language we use is constantly changing, and in education, the words and phrases we use to describe students and student learning change often. Sometimes this reflects changes to what we know in the learning sciences, cognition, and psychology—for example, how educational psychology and special education moved from talking about "mental retardation" in the 1970s to "learning disabilities" today. Sometimes these changes come from policy and legislation, like when we talk about students with IEPs or 504 plans, both of which are terms written into federal law. Much of the time, these changes to language are outside our sphere of influence. We rarely are the ones who initiate such a change! But as we know from the previous two chapters, the language we use matters.

In this exercise, we'd like you to spend some time thinking about your own journey with language about students and student learning. Table 4.1 has several terms that we use to describe groups of students, student performance, or other characteristics that relate to our work as teachers.

■ ■ ■ Try This

For three or four of the rows in Table 4.1:

- identify the characteristics of this group of students who use this term,

- describe other terms that you've used or you've heard used for the group, and

- think about how each term relates to an asset-based perspective on learning. ■

Table 4.1 will get you started with this work, and we've illustrated an example there. You might not want to work with every term in the table—perhaps pick three or four to start. You can revisit this activity multiple times. We have also included a few blank lines on the table to add some terms that may be important to you and your colleagues that we did not consider. The goal for this activity is both to acknowledge the ways in which our language has changed and evolved and to think together about how different pieces of language relate to asset-based perspectives.

When you describe other terms that you've used before, this step might include terms that are no longer considered appropriate; if you are working with colleagues, set some norms for whether you want to raise these terms and how to discuss them if you do.

TABLE 4.1 Exploring Terms for Groups of Students

CURRENT TERM USED	CHARACTERISTICS OF THIS GROUP	OTHER TERMS YOU'VE USED TO DESCRIBE THIS POPULATION	HOW DO THESE TERMS REFLECT AN ASSET-BASED PERSPECTIVE?
504 students	These are students who qualify for classroom accommodations because of a disability but do not qualify for special education services under an IEP. This includes a very wide range of disabilities.	Border students, bubble students, behavior-challenged students	This designation categorizes students by a section of the federal Rehabilitation Act of 1973. It's procedural in that it tells us what our legal requirements are but doesn't give us any information as a group label about students, their challenges, or the assets they bring to the classroom.
IEP students under the Individuals with Disabilities Education Act (IDEA) of 1990 (reauthorized 2004)			
Serious Emotional Disability (SED) students			
Neurodiverse students (broadly and inclusively defined)			

TABLE 4.1 *(Continued)*

CURRENT TERM USED	CHARACTERISTICS OF THIS GROUP	OTHER TERMS YOU'VE USED TO DESCRIBE THIS POPULATION	HOW DO THESE TERMS REFLECT AN ASSET-BASED PERSPECTIVE?
Black students			
Latino/Latina/Latinx students			
Indigenous students			
Asian American/ Pacific Islander (AAPI) students			
Caucasian students			
Low socioeconomic status (SES) students			
Free and reduced lunch (FRL) students			

What did you notice as you worked on the Alignment exercise? Several things occurred to us as we were thinking about how we describe groups of students and how we developed this exercise. First, talking about groups of students is challenging—and potentially uncomfortable. On the one hand, there's a need to use descriptors that identify students in different groups and with various characteristics. One the other hand, we need to be thoughtful about preserving students' identities and individuality and not losing them in the characteristics of a group. We also note that within each group, there is a great deal of diversity. Let's think about students with 504 plans as an example. We might have students with a wide variety of characteristics that are addressed by their 504 plans in our classroom—students with Attention Deficit Hyperactivity Disorder (ADHD), students who have experienced trauma, students who are neurodiverse, and students with various learning disabilities whose assessments don't qualify them for special education services under IDEA.

You might have noticed that there's one type of label that we often hear and use that we didn't include in the exercise: labels related to student ability or performance. These can take various forms—discussions of "high," "middle," or "low" students; course-related labels like "honors" or "Advanced Placement" or "AP" students; or descriptors that focus on performance like "quick thinkers." Think back through some of the ideas we talked about in chapters 2 and 3. Why might these labels be particularly problematic as we think about asset-focused language? Bring your thoughts and ideas about these labels into the next sections of this chapter. We'll explore four key ideas: labels, student-first language, language that empowers student ability, and language we use with colleagues.

As we get started, an important contextual note. In general, we want to call groups what they would like to be called. Student-first language *usually* is well harmonized with this goal, but in some cases, groups prefer identity-first language. For example, the Deaf community (who capitalize Deaf for very specific reasons) generally prefers an identity-first collective. We focus in this chapter on student-first language with the acknowledgment that, in some cases, divergence from that choice is the good and right thing to do to respect peoples' wishes.

WHAT'S THE ROLE OF OUR LANGUAGE ABOUT STUDENTS IN FOSTERING ASSET-BASED LANGUAGE?

One note as we dive in: You may read this chapter and notice aspects of the language that you have used, or currently use, of which we are critical from an asset-based perspective. First, our language is constantly changing and evolving. Some of the labels and terms we're going to discuss in this chapter are ones that we have used ourselves. We acknowledge that our language is imperfect and hope that you, like us, are always seeking ways to grow and improve how we talk about our students and their mathematics learning. We also want to note again that some labels and terms that we are pressed into using may not be directly within our control. They may be imposed on us, formally or informally, by our school building, district, state, or nation. Part 3 of this book (chapters 8 and 9) discusses how we can make changes within our systems, and we'll return to that issue when we get to these chapters. For now, our discussion will focus on unpacking and evolving our language and a critical examination of the terms and labels we use with respect to asset-based perspectives.

LABELS

For the purposes of this discussion, we define *labels* as the ways we refer to students, groups of students, and student performance in the context of mathematics teaching and learning. We use labels in and out of the classroom. Some labels are ones that our students have direct access to in that they hear and see them. Other labels are ones that may not be shared directly with students (although they often hear and must make sense of these labels) in that they're more directly used between colleagues, with administrators, or with community members. There are also less formal labels that we use when we're talking about teaching and learning in informal settings like the community, and ones that we hear from others in these contexts.

We frame our discussion of labels in two categories: labels that relate to student characteristics and labels that relate to student performance. These two categories may not be fully distinct, but we find them to be helpful in framing the conversation. Let's think about each category from an asset-based perspective.

LABELS THAT RELATE TO STUDENT CHARACTERISTICS

Labels that relate to student characteristics include labels related to race, ability status, gender and gender identity, and socioeconomic status. These labels are important in the context of teaching and learning as these are demographic categories that we often are required to track. These labels can also serve as important windows into issues of equity—thinking about who has access to what mathematical experiences as, historically, students of some racial backgrounds have not had the same access to high-quality secondary mathematics as their white peers (National Council of Teachers of Mathematics [NCTM], 2018, 2020). Although these labels can be useful for understanding our students and their opportunities to learn, they also bring with them preconceptions and stereotypes about our students' backgrounds, their prior learning, and often their predisposition toward mathematics.

As humans, we as teachers will have stereotypes and preconceptions about labels related to student characteristics. These beliefs are normal and unavoidable as it is a part of the way that

> ## Tip
> When talking about student characteristics, be cautious not to associate all students who have a particular characteristic with a particular achievement profile.

humans think and reason; minimizing or ignoring these predispositions leads to generalizations that are problematic like, "I don't see color" when referring to race and further marginalize groups of students (e.g., C. M. Steele, 1997). Worse, statements like these paper over the lived experiences of the individuals that make up those groups.

As teachers, teaching equitably means that we're not presuming what students can or cannot do based on any set of characteristics. Understanding students' lived experiences can help us as teachers understand their opportunities to learn and tailor our instruction. We also need to be aware of the preconceptions that we might have related to students with specific characteristics and work, individually and collectively, to challenge those preconceptions. Providing space to hear students' lived experiences and how those relate to their opportunities to learn mathematics is a great way to challenge such preconceptions. The NCTM Effective Teaching Practice, *elicit and use evidence of student thinking*, can be a particularly helpful guidepost here. This practice reminds us that we should be providing students opportunities to share their thinking and work to understand it as teachers rather than making presumptions or assumptions based on characteristics or even past academic performance.

LABELS THAT RELATE TO STUDENT PERFORMANCE

We also use labels that describe student performance in a variety of ways. Like student characteristics, sometimes these labels are ones we're required to use in some way by our school, district, or state. Other times, they serve as shorthand for us to describe aspects of student performance.

One category of student performance labels relates to how students are enrolled in courses or have access to particular academic supports: honors students, AP students, special education students, intervention students. A second category includes more general descriptors like high, medium, or low students. We also might describe students based on roles that they play that relate to their performance such as peer tutoring students, homework helpers, early finishers, and so on.

> **Tip**
>
> Use student-first language where appropriate. Shift how you talk about students from
> _____ students (honors students, visual students) to students who _____ (students who take honors classes, students who struggle with symbolic representations).

Labels about student characteristics and labels about student performance can work against an asset-based perspective. Labels like this can serve to broadly describe what students can and can't do. For example, when we say, "Homework helpers, please find a partner and talk with them about the problem," what message about mathematics competence are we sending to the students labeled as *homework helpers*? What message about mathematics competence are we sending to students who are not homework helpers? We also know as teachers that student performance varies from day to day and moment to moment and can change when the math content we're working on shifts. Will today's student labeled as a *high flier* when we're working on quadratic functions still be a high flier when we make sense of trigonometric identities tomorrow? The use of labels can obscure student assets both for us as teachers and for our students as learners.

LANGUAGE THAT EMPOWERS OR DIMINISHES STUDENT LEARNING

Mr. Leonard's class was getting started on a lesson related to modeling with functions in his Math 3 class. He wanted to start with some reminders about different families of function to set students up for success in choosing appropriate families to model and to discuss the different features of function families. When his students came in, they saw the following warm-up on the board:

Identify the family for each of the functions below.

1. $f(x) = 3x^2 - 5x + 4$

2. $f(x) = 8x - 3$

3. $f(a) = \sin(8a)$

4. $f(x) = |-2x + 1|$

5. $f(n) = 3^n + 5$

6. $f(x) = x^3 + 11x^2 - 9x + 7$

Mr. Leonard got students started by saying, "This should be quick and easy for you—look at each function and tell me what family it belongs to. If you absolutely need to, your notes from last semester might help."

Many students got started right away, jumping around across the six functions to label them with family names they remembered. Some students could be heard murmuring, "Is exponent a family?" and "I think it's something about degree." A few student hands shot up in the air; Mr. Leonard walked past each, not quite breaking stride but quietly saying, "You should remember this from last semester. Dig into those notes if you need to."

What language did you notice in Mr. Leonard's story that may have empowered or diminished student learning potential? Where might each of his teaching moves fall on the deficit-to-asset continuum?

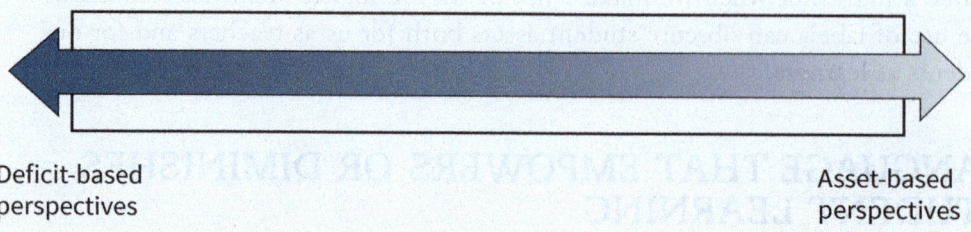

Deficit-based
perspectives

Asset-based
perspectives

Mr. Leonard's warm-up was definitely well intentioned, as was the language that he used with students. His hope was that this warm-up would provide a quick review and introduction to language that would be used in the lesson. He knew that the terms that he was looking for were ones to which students had been exposed earlier in the year. He wanted them to use the resources that they had to help if they didn't immediately know the answers. And he wanted the work to take a short amount of time. As a teacher, we can understand his aims in how he framed the warm-up task.

That said, there are a few pieces of language that may send unhelpful, or even damaging, messages about student learning and student performance. He talks about the task as *quick and easy*. How might a student who struggles with memorization respond to this? He adds that students should use their notes if they *absolutely need to*. Thinking again about a student who may struggle with immediate recall, how has this comment positioned the student? Saying that students should remember this is again well-intentioned as it flags that this is not a new topic, but it also may suggest to students that if they don't remember, they have somehow failed. These messages in a warm-up activity could lead to disengagement for the remainder of the lesson along with the longer-term impact on students' identities and their conceptions of themselves as mathematics learners.

> **Tip**
>
> Shift from language that communicates how easy or difficult a task should be to language that bounds the amount of time you'd like students to spend on the task instead.

Let's consider a different version of Mr. Leonard's launch.

Mr. Leonard's class was getting started on a lesson related to modeling with functions in his Math 3 class. He wanted to start with some reminders about different families of functions to set students up for success in choosing appropriate families to model and to discuss the different features of function families. When his students came in, they saw the following warm-up on the board:

Identify the family for each of the functions below.

1. $f(x) = 3x^2 - 5x + 4$

2. $f(x) = 8x - 3$

3. $f(a) = \sin(8a)$

4. $f(x) = |{-2x + 1}|$

5. $f(n) = 3^n + 5$

6. $f(x) = x^3 + 11x^2 - 9x + 7$

Mr. Leonard clapped his hands and said, "Okay, let's get rolling. I'd like you to take no more than 5 minutes on this task. Some of these you might know right off and some you might need to look up. Notes from last semester, a shoulder partner, or the wide world of the Internet are good resources to support you. Let's share the what and the why for these 281 seconds from now."

This revised version of the warm-up has the same goal: a fast review that allows students the opportunity to recall some key ideas that will be built on in the lesson to come. But the small shifts to the language here are important. Mr. Leonard gives a clear time frame for the work and provides a short set of resources that students can use to generate the function family names. He explicitly notes that you might remember these immediately but that's not necessary, nor is quick recall the goal. If the goal here is for students to share the family and how it matches the function, it likely doesn't matter if students gather that information from memory, from a partner, or from an outside resource. This way of framing the activity provides students with implicit and explicit opportunities to leverage a variety of assets.

LANGUAGE WHEN WE COMMUNICATE WITH COLLEAGUES

■ ■ ■ Try This

Read this interchange between teachers in a professional learning community (PLC) meeting discussing their high school geometry course. What sounds familiar about this exchange? What aspects of asset-based language are evident? What aspects of deficit-based language are evident? ■

Ms. B: Hey, did you try that task from Illustrative Mathematics that we found to start out the unit on dilations? I'm supposed to start it tomorrow with my Regular class.

Mr. J: Yeah—I did it yesterday with the Honors kids. It went well—they had lots of great insights during the exploration. We're a day behind in the Concepts class so we're doing it tomorrow too.

Mrs. N: I'm worried about my Regular and Concepts kids. How long did you give the Honors class to work in small groups?

Mr. J: Probably about 20 minutes or so. We did a whole-class discussion of the solutions after.

Ms. B: I'm really looking forward to that—I want to use that as a springboard to get to the formal definition of a dilation.

Mrs. N: That would work great in Honors, but I don't think the Regular or Concepts class is going to be able to use the formal language to get to that definition, Ms. B. You might have to present them or give them some hints.

Mr. J: I might trim down the small-group work to 15 minutes, Ms. B. The Regular class can get off topic quickly.

What did you notice in that interaction? As educators, this sort of exchange sounds very familiar to us, and it comes from a good-hearted place. These three teachers are thinking about their students' learning, how to best structure their teaching of the mathematics content, and are receptive to learning from one another and sharing their experiences. We also see a desire for important mathematical ideas to come from students, as evidenced with Ms. B's comment about using student solutions as a springboard for the formal definition of dilation. All three teachers are committed to having their students discuss and explore important mathematics. These classrooms are all likely places in which students will have the opportunity to bring mathematical assets to the table.

A few elements of language in the interchange, however, might fall more toward the deficit side of the continuum. First, we see students broadly discussed using labels related to the classes that they're in (Honors, Regular, Concepts). Although the vignette doesn't give us insights into the similarities and differences in each of the three courses, we've noted in the previous section that labeling students based on course names that are associated with achievement levels can be problematic. What images come to mind for Ms. B., Mr. J, and Mrs. N about their students when they hear Honors, Regular, and Concepts? What images come to mind for you?

> **Tip**
>
> Separate actions that redirect student behavior from structuring students' opportunities to learn. Limiting time to discuss and consider mathematical ideas shouldn't be a response to off-task behavior.

The discussion about the Regular and Concepts students perhaps not being able to see important ideas related to the definition of dilation is an example of deficit framing particularly because there is also a suggestion to cut down the time students have to discuss and explore at the end of the story. But let's be clear here—these are not unfounded concerns! As teachers, we want students to have the best possible opportunity to develop new math understandings. We see these teachers thinking hard about how best to accomplish that with students. And off-task behavior is certainly a concern—we have all experienced it, and we've all had moments when winding down student conversations has been one tool we've used to end that behavior. But is preemptively cutting down the time students have to discuss a productive approach? Is it equitable to the students in the Regular and Concepts courses? These are important issues to consider as we seek to allow students to surface and leverage their language-based assets in math class!

When we hear these sorts of comments, we can use three strategies to help us pause and consider language use. These three strategies turn the focus on the language and can be excellent tools to enter into a discussion of language that doesn't feel accusatory or blame filled but invites a deeper discussion of meaning.

The following three strategies can be used to help address deficit-based language:

1. Ask the speaker to **unpack** the language used to help understand the speaker's intent and dive more deeply into why they chose that language. For example, you might ask someone to say more about what they mean by "high flier" students.

2. **Disrupt** the language by restating it and identifying it directly as deficit-based language. You might follow that with inviting the opportunity to revise that language.

3. **Redirect** the conversation from the label to the ideas behind the label. For example, if a student mentions that they reached the answer before other members of their group, you might redirect the conversation by asking about the thinking behind their response or what other members of the group helped them understand in the discussion.

■ ■ ■ **Try This**

Revisit the PLC interchange. Identify one place you would *disrupt* one of the teachers' deficit-based language and describe what you would say to disrupt that language. Identify another place in which you would *redirect* the conversation to a more asset-based approach to student learning and what you would say to redirect the conversation. ■

Shifting and disrupting language between colleagues is hard! It can be uncomfortable for many reasons (see the Digging Deeper feature in this chapter for more on this topic). But this work is important to build a shared understanding of how we can better support asset-focused teaching and learning. If we don't find ways to disrupt deficit-based language and redirect the focus of conversations to students' assets, we tacitly accept that it's okay to express deficit-based views about students and their learning. So let's look at a couple of places we could disrupt and redirect the previous story.

> *Disrupting Mrs. N's characterization: "That would work great in Honors, but I don't think the Regular or Concepts class is going to be able to use the formal language to get to that definition, Ms. B. You might have to present them or give them some hints."*

Tip

Open prompts like, "Say more about that" can help surface underlying beliefs about student learning and student capability in nonthreatening ways.

This excerpt from the conversation paints the students in the Regular and Concepts class as less capable of using formal math language or seeing important math ideas. This example shows the most obvious use of deficit-based language and is a good choice to disrupt. Ms. B might choose some different ways to respond that cuts off the deficit-based language and shifts the view towards student assets. For example, Ms. B might say, "I've found that the students I have in those classes do a great job of explaining math ideas. We sometimes have to tack the formal language on as we go, but I sometimes see that in my Honors class too. The ideas are what's most important."

> **Redirecting Mr. J's suggestion:** *"I might trim down the small-group work to 15 minutes, Ms. B. The Regular class can get off topic quickly."*

Let's assume positive intent here. Mr. J may be absolutely right about students in this class getting off topic, and he's making a kind-hearted suggestion to Ms. B about what she might consider doing. But as we noted, curtailing the time for students to work may not be an equitable strategy. Ms. B could consider redirecting the conversation here to focus on productive student scaffolding strategies. For example, Ms. B might say, "Actually, I think I want to give these students as much time, if not more, to try to explore and see the connections. I'm trying to think about some interim questions I could ask small groups to provide some check-ins on progress. Can I bounce some of my ideas off of you?" This move shifts the conversation to a teaching strategy that's likely to support student assets and recruits Mr. J as a thought partner in considering how to scaffold the learning that Ms. B would like to see with her classes.

■ ■ ■ Digging Deeper
The Nuances of Language and Colleagues

Recognizing language that may frame students in a deficit-based way is the easy part. Making changes to that language, however, both for ourselves and within a school community, is hard. This discussion aims to identify some of the key challenges involved in making changes to that language, particularly when we're interacting with others. We also provide a few suggestions for ways to mark deficit-based language and encourage dialogue and reflection that promotes shifts to asset-based language.

As we work through this section, consider two contexts in which language that may be deficit-based can occur: in the classroom (particularly from students) and in conversation with colleagues. We address both as appropriate. ■

SYSTEMIC CONSTRAINTS

The constraints of our teaching and learning systems may impose language that can be deficit based on us. Earlier in the chapter, we considered the idea of course labels that imply differing levels of prior achievement as language that can serve to label students. Schools may require us to use demographic categories that may not match contemporary language (for example, the U.S. Census continues to designate American Indian as a demographic category as compared with more specific or other more contemporary terms).

When we are limited by systemic constraints, the messages that we send within those structures are important. For example, if we have an Honors or Advanced section of a course, it's important to recognize publicly the framing that this label might have. A teacher might start a year explicitly discussing how the class is labeled, going on to note that we all have diverse ways of thinking and it's important for us to listen to and hear one another. A teacher might also note that all students have times when they can solve a problem quickly and easily and other times when they might struggle, acknowledging these differences and the value of the class's collective assets in moving everyone forward. A class labeled with a term that implicates prior struggles (like *Concepts* or *Fundamentals*) could be provided with similar messages to highlight the assets that students bring.

One means to address this issue with colleagues is to refer to individual students and groups of students in ways that don't use those course labels. For example, I might use specific student names when I'm sharing examples of thinking that took place in class. When talking about a specific course, I might discuss my third period Geometry class rather than saying Honors Geometry. These small shifts can make a big impact in culture and ultimately lead to systemic conversations about how to label courses and group students.

POWER RELATIONSHIPS, PERSONAL RELATIONSHIPS, AND LANGUAGE

As egalitarian as we wish to be, embedded power relationships exist all across our education systems. As teachers and responsible adults, we hold a power differential with respect to our students. As colleagues, issues of seniority and leadership (among others) can create power differentials of all sorts, with administrative roles constituting another layer.

When we disrupt deficit-based language with students, we must always take particular care that we do so in ways that aren't perceived as critical and that are productive and thoughtful while still being corrective. Just as we would be thoughtful about how to discuss student errors publicly, we should also make sure to consider the approach we use to disrupt language. For example, if we overhear a student boasting about how quickly they solved a problem, we might not want to call that student out on the spot. Instead, we might have a general conversation about speed as compared with thinking at the close of the class period. In some cases, we may also want to ask students to share how a label made them feel in terms of their competence. This can apply to deficit-based language that gets used but perhaps more powerfully, highlighting asset-based language that empowered students can be a productive strategy for addressing the power relationships.

Power relationships among colleagues are more challenging. A productive way to deal with these situations is to set ground rules for talking about students (and teachers!) during the start of each academic year. A set of agreements about using asset-based approaches can be helpful as they give us some norms to point toward when we're disrupting language. For example, a PLC might agree to use student-first language whenever possible and post that as a norm in their meeting space. When a teacher might use deficit-based language rather than student-first language, one can gently redirect and note that the rephrasing reflects that specific norm in action.

For both students and colleagues, our personal relationships are also in play when disrupting and redirecting deficit-based language. After one of these instances, it's helpful to check in after a day or so and make sure there aren't lingering bruised feelings. This suggestion may seem like an odd inclusion in this section, but these brief check-ins themselves send an asset-based message—that we're interested in knowing how people made sense of an interaction and hearing their thinking. It helps to model and reinforce the same asset-based ideas we're trying to enact in our teaching.

Reflect, Apply, Transform

We've looked at several stories in this chapter from classrooms and professional learning communities and have considered how the intended and unintended messages might be related to our language use. We've talked about best practices regarding student-first language and ways that our messages about student learning can support an asset-based perspective or perpetuate deficit-based perspectives. We've discussed how we might disrupt and redirect language that perpetuates deficit-based perspectives with our colleagues and with our students. The situations in this chapter were hypothetical, but we certainly found them to be emblematic of conversations to which we have been a party during our years of teaching and working with teachers. We hope they rang just as true for you.

Now we'd like to challenge you to change those hypotheticals into reality. We invite you to record a lesson or a PLC session (make sure you get appropriate consent from students and their caregivers or colleagues as appropriate). Then listen to the recording to note language that reflects asset-based and deficit-based perspectives. (You may also wish to use an automatic transcription mechanism available in popular word processing programs to generate a transcript.) If you're working on a lesson on your own, make some notes about language you may want to shift and integrate those shifts into your planning practice. If you're working with colleagues, consider analyzing the language together and making a plan for how to shift language going forward. This collaboration could include creating shared back-pocket disruption language that is agreed on and all team members can use without judgment when they hear deficit-based language.

Grace is an important consideration in this work. Language change is difficult, and we always should assume positive intent—we are all trying to work to support stronger student learning and to enact stronger teaching practices and making an effort to hold our students' potential in as high regard as possible. It's also important to be kind to yourself and to your colleagues as you're making changes and understanding that change is not a linear process.

1. What are aspects of language that you've heard in your building or district, or used yourself, you'd like to change?

2. How do students make sense of the labels used in the school and district with respect to their math identities?

3. In the past, how have you disrupted deficit language that you hear in your school or district?

PART 2

Asset-Based Routines

In Part 1, we examined how asset-based language impacts outcomes for students and identified strategies that can transform our classrooms and our schools. In Part 2, we will discuss various routines we use in the classroom to deepen our understanding of how these routines fall along the deficit-to-asset continuum. Throughout the three chapters on asset-based routines, we are encouraged to reflect individually and collectively to answer these questions: Which routines are more asset versus deficit based? What actions can be taken to transform my classroom to be more asset based to create more positive outcomes for students? We will look at the following three different types of routines:

- Common, day-to-day structural routines

- Engagement routines that center on discourse

- Groups of routines that become practices

CHAPTER 5

What Drives Student-Centered Instruction

The goals of this chapter are to show how the implementation of some of our most common structural routines in the classroom fall on a deficit-to-asset continuum and to make adjustments toward the assets-based approach that enables student-centered learning. A structural routine refers to an action students understand as part of a daily or common activity during a class period or lesson. We will explore aspects of the products of our structure routines that are useful and helpful to our students. We will also examine what may be unintentionally harmful. The aim is to identify asset versus deficit beliefs for each segment of instruction.

> Miles: "When will I know I'm ready?"
>
> Peter: "You won't. It's a leap of faith."
>
> (Spiderman, Into the Spiderverse)

Questions to Consider

- What structural routines foster student-centered learning?
- What structural routines unintentionally suppress student-centered learning?
- How do these routines in current practice impact students' feelings about the math they are learning?

Recently, my spouse and I (Joleigh) planned a staycation. We had seven days to do what we wanted, but we also decided it was important to relax and rejuvenate. We created a list of the things we wanted to do, which involved the following:

- doing something that someone vacationing in our area would do that we haven't done ourselves,

- exploring and finding something fun to do that is farther away than our regular activities but close enough that we could leave and return within the day (part of staycation for us meant we slept at home),

- finding time to relax (our primary goal!), and

- doing something in our house that we wanted to do but never seem to find the time for (not work related!).

With those parameters in mind, we created our list! We scheduled a few touristy event activities and found some hiking spots that would be new to us. The hikes were about 30 minutes away, making them less convenient than our regular go-to hikes. We also decided that our home (not work) project would be to look at photos and organize our photo library. For relaxing, we decided to do this on days one, six, seven, and throughout the week when we didn't have plans. This was the most important part of our staycation, so we decided to ensure this would be a significant amount of our week. We imagined the time we would spend relaxing would be about 40% of our staycation. I selected a nonfiction, noneducation book to read for pleasure. It would be the first in a long time. I was excited!

Our vacation began, and on the first day, we splurged with a massage, then went home and watched a movie, ate popcorn, and planned for the week ahead. I didn't crack the book that day, but I felt it was ok because we still had six days left. I had plenty of time. Throughout the week, we enjoyed our time and experienced where we live in a different light. We hiked new hikes, attended a play, looked through old photos, and more. Time flew by, and once the days for relaxing arrived, we were ready for them! Unfortunately, I relaxed when day six hit (the weekend), but it was more like any other weekend. My vision of curling up on my cozy chair to read for hours did not happen. Instead, we realized our staycation was ending, Monday was around the corner, and our regular weekend routines minimized our time to relax.

Was the staycation a success?

According to our checklist, mostly yes. We felt good about our experiences, but I made the following observation when I reflected. Our primary goal was to relax, but we spent more time doing other activities than we planned instead of just *being*. My goal was just to sit and read, yet it was placed into the schedule after other items were planned—positioned at the end of the week or added here and there for when we found extra time. (Sound familiar?)

TRANSFORM YOUR MATH CLASS USING ASSET-BASED TEACHING FOR GRADES 6-12

The staycation was a new experience; however, the feeling of overplanning and not meeting my goals was not new to me. I have had my share of running out of time in class and not getting to the main goals I planned. I have found myself wondering how the school year flew by, how the time I planned for a unit took more time than originally scheduled, how an assembly (or other school activity) pushed my entire schedule out of whack, and how a class period sped by when the best-laid plans of students engaged in discourse ended before it felt like it even began. The most common experiences that seem to soak up extra time include spending more time on homework than planned, teaching prerequisite skills that we feel our students don't know, and getting students to participate or even start working on a task independently.

ALIGNMENT EXERCISE: PROPORTIONAL REASONING

This Alignment exercise comes from an observation and interview in Mr. Pehrson's seventh-grade class. Mr. Pehrson is an exceptional teacher. His students describe him as fun and creative and say he values their ideas and contributions to class discussions. When reading the observation below, feel free to annotate areas that send messages to students about the goals of the lesson and what is important (or not important) in class.

Mr. Pehrson has planned a lesson on students developing an understanding of proportional reasoning. He is looking forward to engaging students in a task from his curriculum that has students explore qualities of similar rectangles using strategies and representations of their choice. When the bell rings, the class starts with students working on the daily warm-up while Mr. Pehrson moves around the room, passing out papers students turned in the day before. He then scans the room and quickly takes attendance. The warm-up is intentional and includes problems that access background knowledge, plus a couple of quick review problems from the prior unit so that students continue practicing important concepts. Once complete, the class reviews the answers and then discusses homework questions. Although this process seems to move fairly quickly, Mr. Pehrson looks at his watch and grows anxious. Half of the time for class is already gone, and now there is not enough time for students to fully engage in the task of the day, or what Mr. Pehrson told me was the plan for the heart of the lesson.

He makes the decision to launch the task and have students work independently even though there is less time left than he hoped for before the bell to end class would ring. He passes out envelopes of rectangles to each student and provides directions: "Take the rectangles out of the envelope and determine which rectangles are similar. Explain how you determined that they are similar using multiple representations."

For the last ten minutes of class, students move their ten rectangles (labeled "a" through "j") around and determine which are similar to one another. Mr. Pehrson walks around the room supporting a couple of students in initiating the task and observing students in general to see how they are making sense of the task. Two minutes before the bell rings, Mr. Pehrson tells students to make sure they note which rectangles they identify as similar, place the rectangles back into the envelope, and then pass the envelope to the center of the room. The bell rings, and Mr. Pehrson assures them they will pick up the rest of the task tomorrow. He smiles, tells the whole class, "Nice work today!", and as students walk out of the room adds, "I wonder what qualities are important in a proportional relationship?"

■ ■ ■ Try This

Think about these questions yourself, or if you're working with a group, discuss these questions with your professional learning community (PLC) or other colleagues. Once you've done so, continue reading, and we'll share some thoughts of our own related to these questions.

In the vignette that you read about Mr. Pehrson . . .

- Have you ever experienced something similar?

- How do the structures of this lesson and the routines implemented reflect beliefs about student learning?

- What parts of this lesson positioned students to be successful, confident *doers* of mathematics?

- What actions positioned students as capable?

- What routines are most important to you when enacting a typical lesson? ■

I have experienced situations similar to this many times in my career. It used to be that I wondered how to speed up various activities to get to the heart of the lesson. It took years of reflection and professional learning before I started considering that it wasn't about how fast or slow I was going. Once I shifted my focus to align with my belief statement about students, I changed my actions and how I implemented different routines. My belief statement is as follows: All students bring knowledge and experiences to our classroom and are capable of being successful, confident *doers* of mathematics. Let's look more closely at our most common structural routines and consider how we can transform our daily schedule into a more asset-based learning environment that aligns with your beliefs.

WHAT'S THE ROLE OF STRUCTURAL ROUTINES IN FOSTERING ASSETS-BASED LEARNING?

Structural routines are the various parts of the lesson that shape the flow of what students and teachers do in a class period regularly. Although the flow does not have to be the same chronologically from day to day, when a structural routine is being implemented, students understand the expectations and know what to do. The teacher identifies structural routines and are most effective when they share the intentions and reasoning behind the routine with students. When implemented as intended, structural routines are what make the class run like a well-oiled machine. Classrooms lacking clarity with structural routines can look chaotic and frustrating, with students sometimes unsure as to what they should currently be doing and with whom they should be doing it (Leinhardt, 2001; Leinhardt & Steele, 2005). For many years, I (Joleigh) used several structural routines simply because it was part of my training and/or what my colleagues did. When asked, I could also justify why I used them, but hardly ever did I think about how my implementation of these routines sent unspoken (and unintended) messages that sometimes conflicted with my beliefs about mathematics. When I speak with colleagues about this, our conversations often end up concluding that there is never enough time to get to everything. This ultimately led to two main questions about structural routines and where they fall on the deficit-to-asset continuum. How much time is appropriate for different structural routines? How does the implementation of the routines impact student outcomes?

We're going to discuss some of the most common structural routines used in classrooms (e.g., warm-ups retrieval, homework, instructional routines, assessment, and exit tickets) and consider our current implementation to determine where they fall on the deficit-to-asset continuum by asking the following questions:

- How are the products of our activities useful/helpful to students?

- What unspoken messages do we send?

- Do these activities promote a sense of belonging, or do they unintentionally send messages about mathematics that conflict with our beliefs about who is capable of doing mathematics?

Structural routines implemented from an asset-based perspective intentionally position students for success, support their growth in responsible decision-making, and build student agency and identity. Structural routines implemented from a deficit-based perspective often send messages about math and students that are in conflict with our beliefs about students and their learning. Take a look at Table 5.1, which lists some common structural routines, with descriptions.

TABLE 5.1   Structural Routines and Their Descriptions

STRUCTURAL ROUTINES	DESCRIPTION
Warm-ups	Beginning of class activity
Homework or Practice Sets	Independent practice problems
Instruction	Pedagogical decisions of teaching approaches and how they affect learners
Exit Ticket	End of class activity
Assessment during instruction	Informal understanding of student learning Focus on what students know (listening to students) vs. do not know (asking funneling questions to generate a specific response that lets the student know they should think like the person asking the question)
Assessment after instruction: quiz, test	Formal assessment of student learning
Student self-assessment	Student identifies areas of strength and areas of needed improvement

Let's take some time to discuss each structural routine from the table. We will identify attributes that motivate and engage students and recognize attributes that unintentionally reduce a student's sense of belonging or being capable of the work.

SOURCE: iStock.com/SDI Productions

WARM-UP (ALSO KNOWN AS STARTERS, BELL RINGERS, OR RETRIEVAL PROBLEMS)

Warm-ups usually occur at the beginning of class and have a wide range of implementation (see Table 5.2 for examples).

TABLE 5.2 Attributes of Warm-Ups Across the Deficit-to-Asset Continuum

ASSET-BASED WARM-UPS	DEFICIT-BASED WARM-UPS
• Position students for success • Intentionally connect prior learning to new concepts • Access student's funds of knowledge • Provide students opportunities to make decisions and self-assess their understanding • Incorporate opportunities for students to identify strengths in themselves (and others)	• Position some students as being more capable than others • Lack intention and are meant to keep students busy • Focus on answers • Indicate that math is about being fast • Send messages that may shame students who do not complete homework

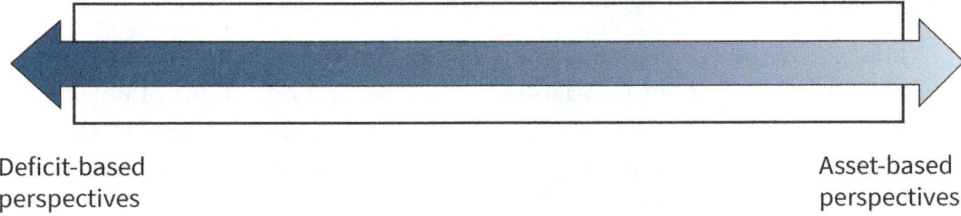

Deficit-based
perspectives

Asset-based
perspectives

Let's look at a few examples of warm-ups. For each, identify the characteristics that place them on the asset-based side of the continuum.

Assets-Based Warm-ups

Example 1: Algebra 1 or Secondary Math I. The class has been working on functions for multiple days:

The teacher begins class by showing students a graph/equation match of five linear functions. Students are to match the graph and corresponding equation. The problems are intentionally similar but different. For example, two have the same slope but different y-intercepts and two are the same equation but with opposite slopes. After a couple of minutes, the teacher asks students to put their pencils down and then provides the answers to the graph/equation match. The teacher asks students to share their strategies for knowing which equation matched a graph. Students experienced linear functions in eighth grade but are now comparing them with exponentials and expanding on their knowledge in this class. The intention of the warm-up is for students to efficiently identify the parameters of the equation and match them to the graph (or to notice aspects of the graph and match it to the equation).

Example 2: Algebra 1 or Secondary Math I: The class is halfway through a unit in which they have experienced characteristics of linear and exponential functions:

The teacher begins class by having students independently write or draw what they know about linear functions and then compare them with what they know about exponential functions. After a few minutes, the teacher asks students to share their responses with a partner. Each student shares their understanding and listens to their partner's contribution. Together, they add new information to their own paper and critique the reasoning of the information that was shared.

Example 3: Geometry or Secondary Math II: The class is halfway through a unit on understanding transformations:

The teacher begins class by having students independently write or draw what they know about translations, rotations, and reflections of objects. After an appropriate amount of time, the teacher asks students to share their responses with a partner. Each student shares their understanding and listens to their partner's contribution. Together, they add new information to their own paper and critique the reasoning of the information that was shared. Students complete a brief reflection self-assessing their understanding of these concepts, recognizing their strengths, and identifying areas where they are still grappling with their understanding.

Example 4: General:

The teacher displays a set of emojis (similar to those shown below) and asks students to choose one that most closely aligns with how they are feeling about the content they are learning (or math class in general). Have them complete the sentence stem: "I chose the ___ emoji because _____."

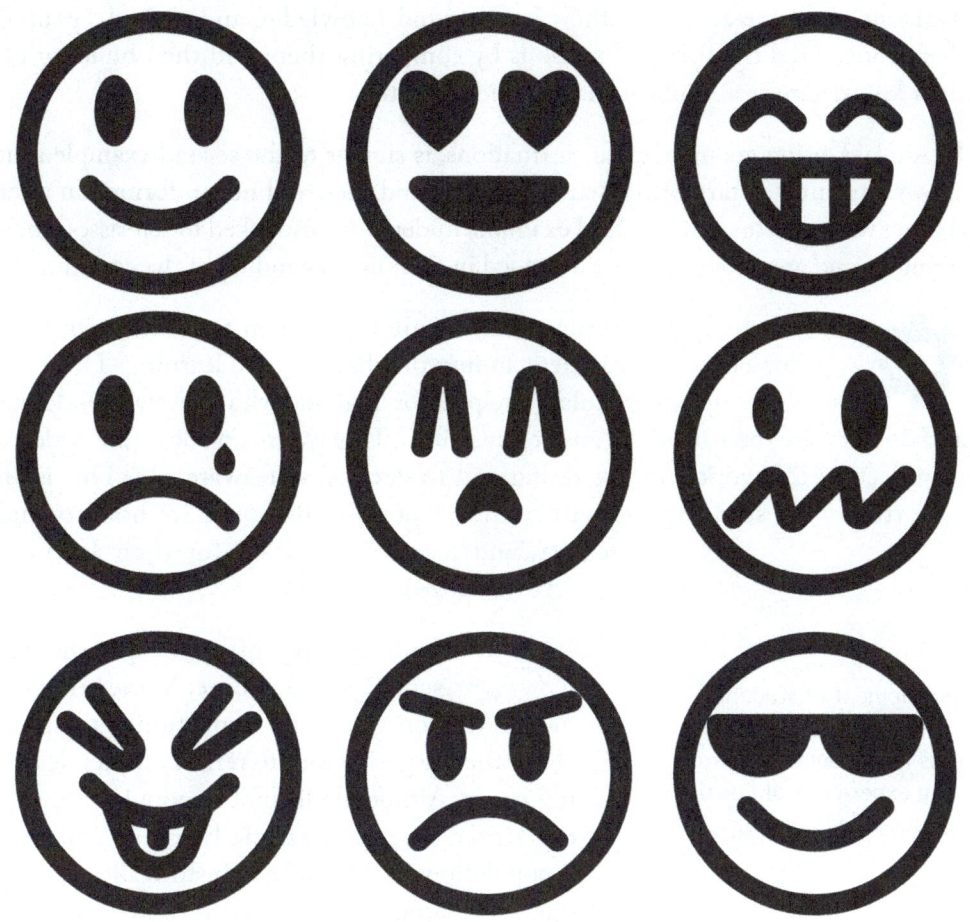

SOURCE: iStock.com/Denis Pobytov

These four warm-up examples focus on what students know, provide opportunities for them to access their background knowledge in an open way, and include interactions that can build onto their current knowledge. Students can express themselves through drawings, in writing, and/or verbally.

In Example 1, students are asked to match graphs and equations. Although students are asked to solve the problems, they are checking their own understanding and not being scored on accuracy. The setting is designed for students to share strategies that focus on their understanding of linear functions, not on whether they got the right answer. Students can self-assess their fluency, and by listening to peers (or contributing to their own understanding), they can expand their knowledge.

In Example 2, students choose what attributes they understand about linear functions and then use that understanding to compare and connect to exponential functions. This warm-up intentionally sets students up to start with what they know. The description of the situation noted that students have had multiple opportunities, over time, to make sense of linear and exponential functions. In this warm-up, they are accessing their background knowledge and then deepening their understanding of both functions by comparing them and then building on their knowledge as a result of sharing with a partner.

Example 3, using geometric transformations, is similar to the second example as it allows students to start with what they know, and then add new information after sharing with a partner. In the third example, students are also asked to self-assess their strengths and areas for growth still needed in their understanding of the concept.

In Example 4, the teacher can monitor and listen to how students are feeling about their understanding of what they are learning. This type of warm-up is particularly helpful for students who otherwise hide out and do not raise their hands to answer questions. This warm-up allows all students to reflect on the work they are doing and to develop self-awareness. This is an opportunity for students to identify their emotions, demonstrate honesty and integrity, and learn to advocate for their learning (CASEL, 2024).

> **Tip**
>
> Support development of student self-awareness by providing opportunities for students to share their learning experience of current content without being judged.

Students have different experiences in school and at home. Some students have not been given the opportunity to speak about their experience or to reflect on their learning needs. Although all student populations may experience this, it is particularly true for students in populations that have been historically marginalized. For example, students with disabilities have often been told specific steps for solving problems, without opportunities to use their own strategies. When they struggle, the intervention in mathematics is usually to slow the learning and go through the steps via repetition. Using a warm-up strategy that promotes self-awareness and self-assessment can shift our understanding of what students know and allows for us to build stronger relationships.

Deficit-Based Warm-Ups

Example 1: Seventh Grade: The class had initial instruction on adding and subtracting integers yesterday:

The teacher begins class with ten random expressions on adding and subtracting integers and tells students they have five minutes to complete the ten problems. If students are late to class, they do not participate in the daily warm-up. The warm-up counts as part of their grade, and the goal is to motivate students to be to class on time.

Example 2: General: The teacher begins class every day with a warm-up and asks students to answer randomly selected questions from the previous day's assignment.

These two examples send unintended and false messages to students about *what* is valued in mathematics and *who* is valued in our class. Early in my career, I (Joleigh) struggled with classroom management and thought part of my job was to teach students how to be "good students." The warm-up was the structure I used to start class and make sure students were in their seats when the bell rang. My primary purpose for the warm-up centered on motivating students to be in class on time and prepared. The questions I asked were often about what I *taught* the day before and not necessarily what the students actually *learned* the day before.

I wanted my students to be critical thinkers and problem solvers. I cared about them deeply and wanted them to feel successful and to love math as much as I did. I didn't realize that some of the messages I was unintentionally sending by asking them to quickly solve problems using procedures they just learned was that math is about being fast, about answer-getting, and that procedures are to be learned in a day after seeing a few examples. Another unspoken message was that if students were late, what they knew or didn't know wasn't important to me because they were not allowed to participate.

This type of structure sheds light on why students think that math is strictly about getting the right answer. We can see why our students, including those with exceptional mathematical promise (National Council of Teachers of Mathematics [NCTM], 2016), or those who "get it" often feel the pressure of needing to get the right answer quickly more than they feel the benefit of learning concepts deeply when structural routines send the message that what matters most is answer getting.

■ ■ ■ Try This

- What new ideas do you have to include in future warm-ups?

- What are some other examples of asset- or deficit-based warm-ups? What are the qualities that make them asset or deficit based?

- What unspoken and unintentional messages might currently exist in your warm-ups? How might you adjust to send more asset-based messages? ■

On the following page, there are examples of ways warm-ups are implemented. In general, warm-ups are not fully asset based or deficit based but lie somewhere along a continuum. Take a few moments to read the scenarios and identify where each falls in the deficit-to-asset continuum and why—and draw in the arrow accordingly. The purpose of this exercise is to notice how warm-ups impact our classrooms and what messages we send as a result of our choices for implementation.

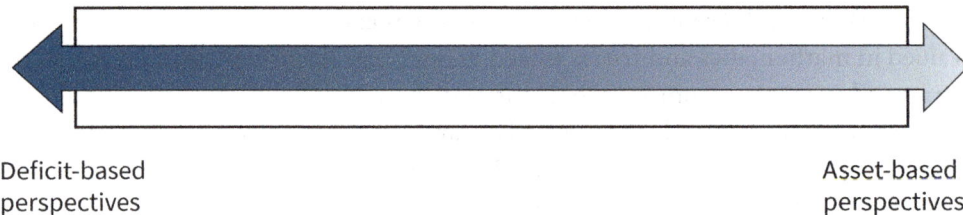

Deficit-based
perspectives

Asset-based
perspectives

Example A: Students work on review problems independently and then trade papers and go over them together in class. Students place a grade on the paper before returning it to the owner of the work.

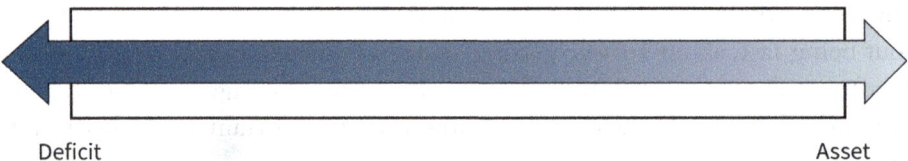

Deficit

Asset

Example B: Students share ways they see the math they are learning in the real world with a partner. They then share why they think this topic is relevant (or not relevant) to them in their lives.

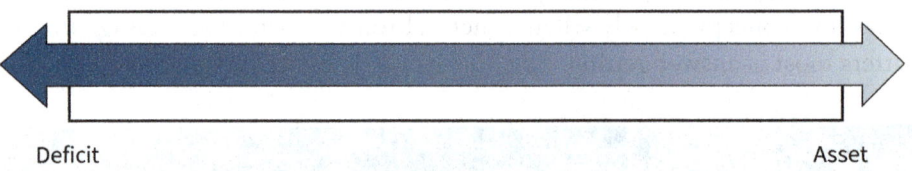

Deficit

Asset

Example C: Students begin class every day with a "Problem of the Day" prompt that has students work on a real-world problem not related to the concept they are learning.

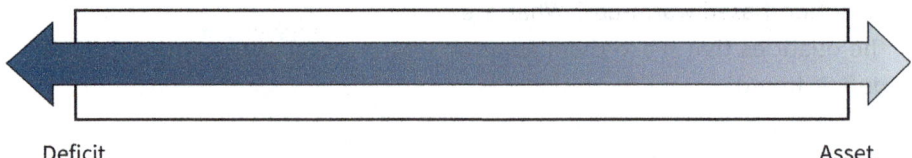

Deficit

Asset

Warm-ups are most students' first daily interaction at the beginning of math class. How do you support students on their learning journey and create an asset-based learning environment (Hunt & Ainslie, 2021) for every student? What are some potential shifts you could make to move your warm-up experiences further along toward the asset-based end of the continuum?

SOURCE: iStock.com/SolStock

HOMEWORK (OR PRACTICE PROBLEM SETS OR CHECK-YOUR-UNDERSTANDING QUESTIONS)

Homework is the collection of problems students complete to check or show their understanding of concepts learned. Traditionally, homework has been an assignment of independent practice problems students are to complete, as the name indicates, at home. The intention of homework seems to be universal on the surface: something students do to master the concepts they are learning in class. However, when we asked teachers to explain further why homework is helpful, we received almost as many different answers as there were teachers. Responses included the following:

- Support students in becoming responsible for managing their time to complete tasks independently

- Practice solving problems to become fluent

- Check their understanding of previously learned concepts

- Boost students' grades

- Encourage students to take good notes in class so they can follow examples when doing problems on their own

- Identify areas where students recognize they need more help (because they didn't understand the explanation as well as they thought they did when they heard the teacher go over it in class).

The many different responses highlight both issues and opportunities with homework. Like warm-ups, this structural routine can cultivate a positive learning environment and empower students, or it can create anxiety and reduce student participation. Like warm-ups, *how* we implement homework makes the difference.

Asset-based learning environments can use Practice Problem Sets to position students for success in their learning journey. The time devoted to reviewing homework in class can be meaningful and discourse based (Otten et al., 2015) if the problems and routines around them are thoughtfully structured. Practice Problem Sets are asset based when they have these attributes (see Table 5.3).

TABLE 5.3   Attributes of Asset-based Practice Problem Sets

ATTRIBUTES	ASSETS
Problems are intentionally selected from concepts in which students already have a certain level of fluency	These problems are intended to build procedural fluency (NCTM, 2023) *after* students have learned the concept. Spacing these problems out over time (Brown et al., 2014; Hattie, 2008) supports students in building confidence and deepening their understanding of important concepts.
Problems are selected to access background knowledge and connect prior and current learning. For example, just before students begin a proportional reasoning unit, they generate equivalent fractions and explain their process for doing so. This approach reinforces their understanding of equivalence while prepping them to understand the relationship of proportional reasoning to equivalent ratios.	Drawing on background knowledge positions students to recognize what they know and make connections supporting reasoning and sense-making about the new mathematics they are learning.
Problems are about quality, not quantity (Terada, 2018). Since this routine is about checking for understanding, focus on quality. In Terada's Edutopia article, the recommendation for time spent on homework from the National Education Association (NEA) is approximately 10 minutes per grade level. That means a seventh grader would spend a total of 70 minutes across their classes. If Math is around one third to one fourth of this time, you would want not to have more than 20 minutes of math homework.	Students can focus on checking their understanding and reflecting on where they are in their learning journey.

ATTRIBUTES	ASSETS
Problems can cover more than one concept. Interleaving problems (using a variety of problems) promotes fluency and problem-solving (Brown et al., 2014; Hattie, 2008). When students only practice one problem type in a practice set, they may mimic or work but not learn.	Students learn math. Interleaving promotes fluency and problem-solving
Homework should be designed *for* the students and not intended to be graded for accuracy by the teacher. Data show students resort to mimicking, cheating, and withdrawing from participation when homework is graded for accuracy (Liljedahl, 2021).	The purpose is for students to feel safe practicing and providing students with feedback for a path toward improvement. Homework (or checking for understanding) problems are about learning from mistakes instead of being penalized for them.
Problems are used as opportunities for discourse.	Instead of honing in on the mechanics of one problem (after another), the focus is on the larger mathematical idea. In addition, opportunities for discourse include students making sense of the generalizations of a topic by looking for structure (Standards for Mathematical Practice [SMP] 7), repeated patterns (SMP 8), or connections (Otten et al., 2015).

Deficit-based practices for homework include focusing on answer-getting and grading for accuracy when students are learning new content. When learning new content, students need time to make sense of the concept. During this time of initial learning, student thinking is fragile. They are making connections with their background knowledge to the new material and adjusting their thinking as they gain more insight over time. During this time, repeatedly sending students home to practice new concepts often leads to repeating the same mistakes. This process is not only confusing for students but also frustrating; if it happens often enough, they stop doing homework. From here, it doesn't take long to disengage. In *Building Thinking Classrooms* (Liljedahl, 2021), traditional homework involving completing many problems similar to what was just learned in class has the following deficit effects:

1. Students consider traditional homework one of their top three stressors.

2. Many students resort to cheating, mimicking, or not doing it.

3. Marking (grading) homework increases the likelihood of the first two statements.

> **Tip**
>
> Use warm-ups and homework problems that have students practice concepts they are familiar with instead of new content. These problems spiral in difficulty over time. Class discussions for these structural routines focus on the concept, not on the computation.

- What does this structural routine look like in your class?

- What are your current goals for this routine, and how do they align with your intentions and intended outcomes?

- What do you want to start doing?

- What do you want to stop doing?

- What are your next steps so your students are more engaged and can self-assess their understanding in a safe environment? ■

Take time to reflect and then intentionally make decisions about how you would like students to check their understanding in ways that promote power in participation.

SOURCE: iStock.com/ake1150sb

INSTRUCTIONAL PHASE OF THE LESSON

At the heart of every lesson plan, our choices for delivering instruction are endless. How do I access background knowledge? When and how will I provide opportunities for discourse? How do I align the goals of the lesson with the goals of the learning progression? When will new vocabulary be introduced, and how?

How will I know every student understands the goal, and what will I do if and when they do not? This entire book could be a list of questions we ask ourselves when planning instruction. We make so many decisions that we are not even aware of all or perhaps even most of them. In this chapter, we will look at asset-based instruction to reflect on how we think about instruction when we are planning to teach a lesson.

Asset-based instruction is complex. In its simplest form, instruction falls further along the asset continuum if the result of the instruction includes the following:

- high rates of student participation among all students,

- opportunities for students to see value in what they are learning and can see themselves as capable *doers* of mathematics, and

- students provided with multiple opportunities, over time, to show conceptual and procedural understanding of the content they are learning.

Although it sounds simple, achieving these three bullets is quite complex.

Likewise, instruction that falls along the deficit continuum would be the opposite of these traits. Instruction that leads to students withdrawing or not participating or that students see as irrelevant (the infamous "When am I ever going to use this?"), or that assumes students will learn important mathematics deeply in one sitting can easily be seen as falling along the deficit end of the deficit-to-asset continuum.

■ ■ ■ Try This

To increase participation, record yourself giving a lesson or ask a colleague to observe and record the following:

- How often are students asked to share their thinking? How is individual student thinking valued?

- Is the focus on student reasoning or answer-getting?

- How are student's strengths recognized? What is the evidence that we know our students and that peers know each other?

- Are some students positioned as more capable than others, or are structures set up to ensure all voices are heard and valued? ■

Asset-based instruction classrooms provide opportunities for students to see value in what they are learning and to see themselves as capable *doers* of mathematics. Students achieve this insight by developing agency (willingness to engage),

Tips

- Increase opportunities for students to see themselves and others as collaborators by using worthwhile tasks with multiple pathways toward a solution.

- Use instructional routines or practices that promote discourse.

- Pose purposeful, open-ended questions and give students independent think time.

- Ask students to share their thoughts with a partner using a structure that ensures each person is provided time to share as a contributor of information and each student is provided time to listen and capture the ideas of others.

- Use instructional routines that position students as collaborators. Examples include Turn and Talk, Two-Minute Talks, Math Language Routines, and Number Talks (see more in Chapter 6).

a positive math identity, and ownership of the mathematics they are learning. This all happens when students are provided with meaningful contexts, have access to mathematics, and are provided opportunities to "contribute to conversations about mathematical ideas that build on others' ideas and have others build on theirs" (Teaching for Robust Understanding Framework, 2023). Throughout instruction, students are provided choices in problem-solving strategies, including representations, and how they will communicate their learning, including writing, drawing, and/or speaking. Chapters 6 and 7 will detail various instructional routines that contribute to student identity and agency and how these routines empower students to see themselves as capable *doers* of mathematics.

Asset-based instruction classrooms incorporate NCTM's Effective Teaching Practices. *Multiple representations* are used to make sense of problems and to deepen understanding of concepts by making connections between representations. Routines or practices that *orchestrate productive discussions elicit evidence of student thinking* and align the lesson with clear *goals*. Teachers can provide students with multiple opportunities, over time, to show conceptual and procedural understanding of the content they are learning. When planning instruction, understanding how students learn and the learning progression of mathematics is essential for students to *build procedural fluency from conceptual understanding*. A framework that provides students with conceptual and procedural understanding, such as the Comprehensive Mathematics Instruction (CMI) Framework (Hendrickson et al., 2008), means that students deeply learn the important mathematical concepts in their grade level using a learning progression. Instruction should provide students with multiple opportunities to understand major concepts deeply. A learning progression starts with developing the concept by surfacing ideas, followed by solidifying specific aspects of the concept, then having students refine their thinking and build fluency through a practice task (Lemon & Hendrickson, 2023). Students are provided customized support and opportunities to self-assess their understanding throughout the learning progression.

TRANSFORM YOUR MATH CLASS USING ASSET-BASED TEACHING FOR GRADES 6-12

Reflect on these questions yourself, and discuss with your PLC:

- How do you focus your instruction on increasing student participation? What additional ideas do you have as a result of reading this section?

- What instructional routines do you currently use to build positive identities for your students? How can you build more opportunities into your instruction that allow students to "walk the walk and talk the talk"?

- What resources do you use to better understand the learning progressions for your grade band, for your course, and for the mathematics concept you teach? How can you move this aspect of your instruction even further along the asset-based instruction continuum? ■

EXIT TICKETS

Exit tickets are usually one of the last structural routines we use in lessons. Asset-based exit tickets allow the student to self-reflect on what they are learning. Such self-reflection shouldn't be graded but used instead as an opportunity for students to communicate and reflect on their learning. As we collect feedback on student understanding, we can adjust instruction based on information we learn from students as a group. We can also provide individual feedback on their progress and guidance for next steps.

> **Tip**
>
> Increase engagement by having transparent conversations with your students about the intention of your routines.

Examples of asset-based exit tickets include those in which:

1. Students self-assess their understanding using success criteria that the teacher shares in advance

2. Students answer reflection questions about what they know already and what they are still working on

3. Students tackle problem(s) you believe they understand before they are assessed for a grade (and potentially surprised by what they know and do not know)

> **Tip**
>
> Use open-ended reflection questions that allow students to explain what they know and describe what they are working on or share what they are confused about.

Deficit-based exit tickets do not provide helpful information to the teacher or the student. Although none of us intend to implement routines that have a negative impact, some things we do with exit tickets may not support learning. An example of a deficit-based exit ticket would be students being graded on accuracy for the content they were just introduced to that day. Even when students listen and take good notes, they need processing time, multiple opportunities or exposures to practice, and formative feedback from us as teachers when learning a concept. An exit ticket that demands immediate mastery provides for none of these opportunities. The expectation is that if they listen and take good notes in class, then they should get it right.

Exit tickets have the potential to transform learning and teaching. Students benefit when they self-assess their understanding of key ideas. If you focus on asset-based exit tickets and allow students to highlight what they know and consider what they are still working on, we as teachers will also benefit because the informal assessment provides insight into what students know individually and collectively as a class.

THE MANY MODALITIES OF ASSESSMENT

Think about your experience with assessments when you were a student.

What emotions do you feel when considering the tests or quizzes you took? Maybe you were stressed and therefore didn't like assessments. You might have experienced anxiety for any number of reasons. You may have felt like you always wanted more time or never seemed to know which concepts the test would focus on. Or maybe you were excited about assessments. Perhaps you enjoyed the competition or were known to be smart because you did well and received recognition. Maybe your experience was something completely different than described here. Whatever your experience has been with assessments as a student, I'm guessing you have feelings about them, as most adults do! Let's look at ways we assess student understanding and identify opportunities to disrupt deficit-based assessment practices and increase asset-based practices.

Summative Assessment

Summative assessments are what most people think about when they hear the word *assessment*. Summative assessments measure outcomes of what students should know at the end of a learning progression, usually as a unit test, semester exam, or end-of-course high-stakes test. In this situation, assessment is a noun. Assessments can be powerful but can create a lot of emotion for students (and teachers alike). Assessment professionals emphasize the importance of understanding the intended purpose of different assessments. They assert that the better a test serves one purpose, the worse it will do with a different purpose (Zavitkovsky, 2022). For example, state-level assessments provide satellite data (Safir & Dugan, 2021) and are intended to be used to identify overarching student learning trends.

If we use these data to identify skill-level information about individual students, we miss the intention of the test and often implement deficit actions as a result.

Examples of shifts that move us further along an asset-based continuum for summative assessments include:

- providing students with retake or redo opportunities,

- shifting to standards-based grading focusing on competency versus points,

- advocating to disrupt the inaccurate use of satellite data, and

- allowing students choice in how they show or explain their understanding of concepts and skills (Safir & Dugan, 2021).

If you already use some of these strategies, the next step may be to look at how these are being implemented. For example, review your current standards-based grading rubrics to assess whether they convey asset language. Note deficit-based language and shift to asset based. Change headers from "Needs Improvement" or "Below Grade Level" to "Emerging" or "Not Yet Proficient." Shift indicators to reflect what students know instead of what they do not know. Student choice becomes important when we start recognizing individual student strengths. The more we understand what it means to be conceptually and procedurally competent in a given concept, the more we recognize various ways students can demonstrate their knowledge, often more robust than a typical end-of-unit test. Examples of assessments include written prompts, observations, interviews, performance assessments, presentations, and more.

When I (Joleigh) first learned about choice, I was skeptical. Secondary mathematics teachers have many students, and I thought, "How could I possibly provide choice for each of my students?" I was encouraged to start small and then look for opportunities to increase ways to incorporate student choice. I began noticing how I was providing choice during instruction and used this as a start for modifying assessments to incorporate more choice. Each person is on their own journey as an educator when it comes to providing students with choice. If you already do this, I applaud your accomplishments and encourage you to continue expanding in this area (move further along that continuum!). If providing a choice is new for you, start small. In fact, start by thinking of ways that you probably already promote choice. Do students engage in problems that have multiple solution paths? Do you already encourage using multiple representations as tools to make sense of and solve problems? An example of this could be students being asked to analyze a context and determine the zeros of a quadratic function. Students make sense of the problem and then create representation(s) of choice. Representations may include algebraically rearranging the form of the equation to highlight the zeros, graphing the function to identify where it crosses the x-axis, creating a table, and using symmetry or other characteristics of quadratics, using the Quadratic Formula, or another representation/problem-solving strategy. This kind of choice promotes student thinking as they determine which representation(s) support them in leading to a solution. By providing this kind of choice, students develop procedural fluency as they practice identifying efficient strategies and become flexible in solving quadratic functions (NCTM, 2023). Be sure to appreciate wherever you are when it comes to providing options and choices for students, and then commit to seeking more opportunities to grow in this area. And give yourself time and space to make shifts that work for you.

My shift toward a more asset-based implementation of summative assessments occurs as my perception and understanding of how to use formative assessments become more effective. Formative assessments occur during the learning process, focus on improvement (rather than on evaluation), and are often informal and low stakes (Baylor University, 2024). My biggest leap to date happened when my perception of assessing students shifted from assessment (the noun) to assess (the verb).

Formative Assessment

The root word of assessment, assess, in Latin (*assidere*) means "to sit by." In an asset-based classroom, we create opportunities for students' understanding of mathematics such that they feel like someone is sitting beside them along their learning journey. In this case, assess is a verb used to capture the informal, ongoing process of understanding what students already know to support their next steps in the learning progression. Informally assessing student learning means implementing Street Data rules that move us from a pedagogy of compliance to embracing a pedagogy of voice. We spend more time focusing on questioning and listening over information dissemination, providing feedback over grades, and recognizing students as competent and capable instead of seeing them as vessels to fill with information (Safir & Dugan, 2021). For more details about shifting to a pedagogy of voice, we encourage you to read *Street Data*.

Just as caregivers support toddlers in walking, we can support students in learning math. When toddlers learn to walk, caregivers use multiple resources to support progress along the way. We provide opportunities for the toddler to stand and build strength and incorporate customized supports like furniture, our fingers, or other items to hold onto while the toddler practices balancing. Caregivers also encourage independent walking and celebrate milestones. Although we may not sit beside the toddler every moment of their progress toward walking, we know how to provide opportunities, use resources, and support progress through feedback. Likewise, as teachers, we create learning environments in which students are provided opportunities to access their background

> **Tip**
>
> Make the learning goals visible by providing learning intentions (what we want students to learn) and success criteria (explicit descriptions of what it looks like when the content is learned; Frey et al., 2019).

knowledge, contribute their ideas, and build on the mathematical ideas of their peers. We provide customized support and solidify the learners' understanding of concepts over time. We support student progress to "walk the walk and talk the talk" of mathematics. And we celebrate milestones along the way.

When Students Self-Assess

Asset-based assessments align with learning goals and are as clear to students as they are to us as educators. Students are more likely to be successful when they can assess their progress toward goals and explain what it looks like when they have learned what they are supposed to learn. In daily lessons, we monitor student interactions and listen for student discourse that informs us about what they know. Time is built into instruction for students to reflect on their learning journey and self-assess their progress toward the goals or success criteria. The more we move along the asset-based continuum and "sit with" students as they learn, the better we recognize student strengths, build on what they already know, and create opportunities for students to explain their understanding of concepts via speaking, writing, or using representations. When we empower students to self-assess, we support their development of student agency as they gain ownership of the mathematics they are learning.

> **Tip**
>
> Make the learning goals visible by providing learning intentions and success criteria or "I can . . ." statements for students in advance of the learning (Frey et al., 2019).

■ ■ ■ Digging Deeper
Recognizing Cultures as Intellectual Resources

For each structural routine we implement in our classroom, we should consider its purpose, why we do it, how much time should be devoted to it, and who it serves. When we review our structural routines, taking the time to consider the students we serve and their cultures makes a big impact as we make decisions about implementation and expectations for these routines. Asset-based routines recognize students' cultural and linguistic differences as intellectual resources rather than as deficits (Ramirez & Celedon-Pattichis, 2012). What does this look like?

Culture plays a role in our routines, whether we realize it or not. If you have ever thought or have heard someone else say, "I must first teach my class how to act or how to be students," you may actually be thinking, "I must first teach the students in my classroom how to adjust to my own cultural norms." I (Joleigh) have heard this statement from peers and have realized that whether I thought this myself or not, I have been guilty of implementing structural routines in a deficit-based manner. My most memorable experience happened in my

(Continued)

12th year of working in education. I was co-teaching with a colleague. and our class was mostly Latina/Latino/Latinx students. We had spent weeks working on strategies to engage students in discourse. One student, Mario, was not only new to our class but also to the school. He was shy but coming out of his shell as his peers encouraged him and he began to feel safe.

In the meantime, we also used warm-ups as a routine to see what students could do independently. So, at the time, my rule (and cultural experience) was that there was no talking at all during this time and if someone did talk, no credit. As you may have guessed, Mario started talking to his neighbor about one of the problems, and my rule was that this meant he did not receive credit for his warm-up. I was so focused on making sure I had rules and held each person accountable that I didn't recognize how quickly I turned our nurturing, safe environment into an experience that made Mario not want to talk in class again. I am sure it also impacted other students, but at the time, I didn't see the result of my reaction. In hindsight, I would have handled the situation differently.

What would you have done? What could have been a more asset-based solution? ∎

Reflect, Apply, Transform

Let's revisit the structural routines discussed in this chapter in Table 5.4. Consider your implementation of these routines, and identify aspects that foster student-centered learning, aspects that suppress student-centered learning, and other reflections you have to transform your classroom into a strong asset-based learning environment.

TABLE 5.4 Student-Centered Learning and Routines

ROUTINES	WHAT ASPECTS OF THIS STRUCTURAL ROUTINE FOSTER STUDENT-CENTERED LEARNING?	WHAT ASPECTS OF THIS STRUCTURAL ROUTINE SUPPRESS STUDENT-CENTERED LEARNING?	WHAT OTHER REFLECTIONS OR COMMENTS DO YOU HAVE TO TRANSFORM YOUR CLASSROOM?
Warm-ups			
Homework			

ROUTINES	WHAT ASPECTS OF THIS STRUCTURAL ROUTINE FOSTER STUDENT-CENTERED LEARNING?	WHAT ASPECTS OF THIS STRUCTURAL ROUTINE SUPPRESS STUDENT-CENTERED LEARNING?	WHAT OTHER REFLECTIONS OR COMMENTS DO YOU HAVE TO TRANSFORM YOUR CLASSROOM?
Instruction			
Exit ticket			
Assessment			

CHAPTER 6

Increasing Student Confidence and Engagement Using Asset-Based Routines

This chapter focuses on enhancing our awareness of ways to increase student confidence and engagement using asset-based instructional routines intentionally connected to developing a student's positive mathematics identity.

> *"Educators who employ effective collaboration . . . greatly amplify student engagement, achievement, and growth, particularly for those learners needing extra support." (Jenkins & Murawski, 2023)*

Questions to Consider

1. What instructional routines contribute to a student's positive mathematics identity?

2. How do asset-based instructional routines cultivate a positive mathematics identity and support deep mathematics learning?

Do you ever think about school when you are definitely not at school? Do you see math everywhere? Or are you often the person a stranger asks to take a group picture using their phone (because you somehow give off the "I am a teacher, and I am here to help" vibe)? I (Joleigh) am very proud that I am an educator, but sometimes, I catch myself when I am not at school and notice the teacher come out in me. Sometimes, it even happens when I am on vacation.

Each year, my family goes on vacation to a national park. Last year, we visited a location where it is recommended to join tours to learn about the surroundings, so my son organized two one-hour tours for our family. During the first tour, the guide had us walk a distance and then said, "Look around. What do you think life was like? What questions do you have?" Throughout the tour, more and more people asked questions. I learned a lot because of the observations and questions asked by the group and the storytelling of the guide who related information based on the group's questions and comments. A couple of times the guide responded that they did not know the answer, but this only enhanced the experience and increased curiosity. Ever being the teacher, I of course related this to the instructional routine: "What do you notice? What do you wonder?" We were on vacation, so I kept this to myself!

The next day, we joined a tour in which the guide started with a brief history of the people who had lived there so many years ago and then posed the question: "What do you think the people ate when they lived here?" Several people replied, and the guide quickly confirmed correct and rejected incorrect answers. Within 20 minutes and six questions into the tour, no one in the group was responding. My son later joked about the awkwardness. He said that our guide was talking to himself the whole time and that it was like the classroom scene from *Ferris Bueller's Day Off* where the teacher asks a question and says, "Anyone? Anyone?" I thought about the funnel-type questions asked by the tour guide that day and how different it felt from our experience the day before. The experience reaffirmed my impression that education is everywhere. Even here, on vacation, the two different approaches of instruction by our guides played out similarly as they would in our classrooms.

We spend a great deal of time thinking about our instructional routines from a teaching standpoint. It's moments like these with the park guides that might help us think about the impact instructional routines have on our students and their developing identities as learners. As we move through this chapter, try to consider both teacher and student perspectives on the routines we discuss.

ALIGNMENT EXERCISE: IDENTIFYING ASSET-BASED INSTRUCTIONAL ROUTINES

The Alignment exercise presents several instructional routines and a short description of each. Use this exercise to consider where these routines fall along the deficit-to-asset continuum and to reflect on the aspects that had you place them there.

■■■ Try This

Keeping in mind that an individual's mathematics identity is about students seeing themselves as belonging and as capable, look at the list of instructional routines and their corresponding purposes in Table 6.1 to identify where you believe each of the listed routines falls along the deficit-to-asset continuum. ■

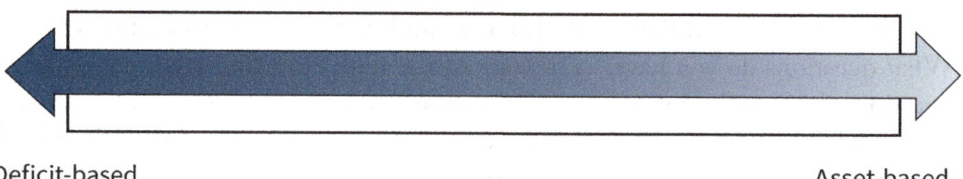

Deficit-based
perspectives

Asset-based
perspectives

TABLE 6.1   Instructional Routines Along the Deficit-to-Asset Continuum

INSTRUCTIONAL ROUTINE	DESCRIPTION OF ROUTINE
Think-Pair-Share	A question or prompt is given to the class. Students are given time to think (and often to write their thoughts down) prior to them discussing their ideas with a partner or small group. Deficit Asset
Number Talks	A 5- to 15-minute classroom conversation around purposefully crafted problems that are solved mentally. Deficit Asset
Two-Minute Talks	Pairs of students taking turns sharing everything they know about a given topic (one minute for each student). Deficit Asset
Notice and Wonder	Students are provided a situation and are first asked, "What do you notice?" Students generate a list of what they notice. After students have shared their observations, the next question asked is "What do you wonder?" Students then share what they are curious about or what they are wondering about in regard to the situation. Deficit Asset
Which One Doesn't Belong?	Students are provided with a few options (usually four, but can have a wider range) and by comparing the options, choose which one is not like the others. Once the student chooses, they also provide a justification for their selection. Deficit Asset
Turn and Talk	Students are asked to turn and talk to a partner about a given prompt. Deficit Asset

INSTRUCTIONAL ROUTINE	DESCRIPTION OF ROUTINE
Math Language Routines	Provide a framework for organizing strategies and special considerations to support students in learning mathematics practices, content, and language. 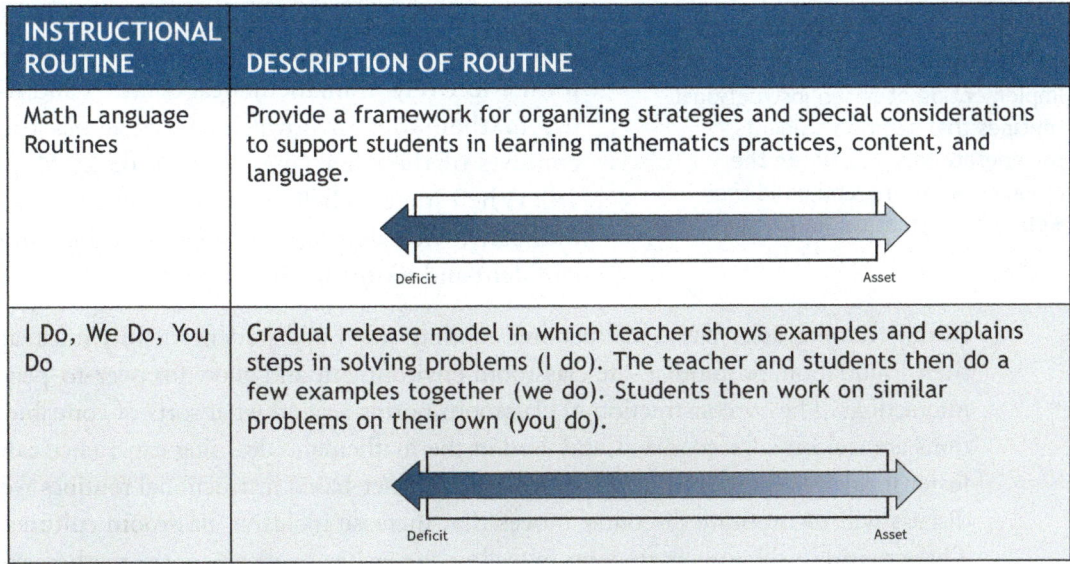
I Do, We Do, You Do	Gradual release model in which teacher shows examples and explains steps in solving problems (I do). The teacher and students then do a few examples together (we do). Students then work on similar problems on their own (you do).

Let's talk about where we placed each routine along the continuum and answer the question "How might the routine promote a positive student identity?" You may choose to use the study guide to capture your ideas by listing the features of the routine that could cultivate a student's positive mathematics identity. Table 6.2 provides an example of what this can look like for the Two-Minute Talk routine. Throughout the chapter, pause and reflect to add additional features of instructional routines that promote a positive student identity.

TABLE 6.2 Features That Cultivate a Student's Positive Mathematics Identity

INSTRUCTIONAL ROUTINE: TWO MINUTE TALKS	
Features:	**How might this feature promote a positive student identity?**
Honor student thinking	*Students use their own understanding to choose what to share.*
Access prior knowledge	*Students recognize what they know and add to their knowledge by listening to a peer.*
Community of learning	*Students develop relationship skills and learn to respect the contributions of their peers.*

WHAT IS THE ROLE OF INSTRUCTIONAL ROUTINES IN FOSTERING ASSET-BASED LEARNING ENVIRONMENTS?

In the first chapter of this book, we aligned a student's positive mathematics identity with students seeing themselves as belonging and recognizing themselves as capable of succeeding in mathematics (National Council of Teachers of Mathematics [NCTM], 2018, p. 25). An individual's mathematics identity is "the

Tip

Implement asset-based instructional routines that position students to generate ideas, listen to the opinions of others, and build their web of mathematical knowledge.

dispositions and deeply held beliefs that students develop about their ability to participate and perform effectively in mathematical contexts and to use mathematics in powerful ways across the contexts of their lives" (Aguirre et al., 2024, p. 14). When students believe in themselves mathematically, they develop the agency to become confident and motivated participants.

In this framing, asset-based instructional routines that build student identity must, at their foundation, promote a safe classroom environment and allow for peer-to-peer interactions. The co-construction of classroom norms around what sorts of contributions are welcomed, recognized, and used in the mathematics learning experience can foster inclusive classroom cultures. Therefore, all asset-based instructional routines we discuss will incorporate discourse moves that increase inclusive classroom cultures. These routines will support students in feeling ownership, or agency, of the mathematics they are learning. Instructional routines promoting an inclusive environment position students so everyone has equal status and value in the math classroom.

When implemented thoughtfully, these routines cultivate an asset-based learning environment. Students feel safe communicating their ideas and taking risks. They feel heard by their peers and their teachers. Students recognize the value of building their own understanding by listening to the ideas of their peers. And they understand that reasoning and problem solving is at the heart of learning mathematics.

SOURCE: iStock.com/monkeybusinessimages

DISCOURSE MOVES

Fostering meaningful mathematics discourse is one of the eight NCTM Effective Teaching Practices and can encompass a wide range of classroom interactions. Discourse moves that accomplish an asset-based learning environment might require a considerable shift in how we, as educators, listen to our students—depending on how far along the asset-based continuum we are at the moment. If we want students to think and reason and for peers to value their contributions, we must train ourselves in how we listen and, more importantly, hear our students. As we learn better to hear our students' ideas and how they think about a problem, we can then align our questions to advance ideas along their solution path. This approach contrasts with asking students questions and listening for them to respond with a specific answer that aligns with a solution path in our mind. If we only know one solution path or way of solving a problem, it is difficult to listen for understanding and easy to try to redirect the student to our way of thinking. When we redirect, students have to shift their thinking. They may get the answer to a problem, but the consequence is that this does not honor the student's approach, which often results in making them think their approach was wrong. This approach also perpetuates the false notion that there is only one way to solve a problem. By shifting the way we listen, we see students as having more knowledge than we may have thought. Giving students the space to think for themselves and develop solutions that make sense to them grows their identity and agency.

For students who learn procedures quickly and/or for students who are identified as having a propensity for mathematics, their belief may be that math is about "answer getting." As a result, the focus is on answers, and time is not spent on digging deeper into understanding generalizations or why their solution works.

Tips

Listen to what students know and truly hear what they say:

- Come from a space of curiosity about their thinking.

- Avoid assuming what they will say or hoping they will answer in the same way you are thinking.

Tips

- Encourage students to explain their thinking and then probe so that they elaborate, modify, or clarify their ideas.

- Allow students to build onto the thinking of their peers, including by making connections to their own work, or by comparing what is similar and what is different with another piece of work, or by critiquing the reasoning of their peers.

- Provide space for students to have independent think time so that all students have an opportunity to develop ideas and that it isn't just the quick processors who appear to be the "better" mathematicians.

- Create an environment where students see mathematics as a process, and encourage students to revise their work and refine their understanding over time (Chapin et al., 2009; Kazemi & Hintz, 2023; Spangler & Wanko, 2017).

For this group of students, asking probing questions and having them clarify their reasoning may be uncomfortable at first. Still, when this norm is established, they come to value their deeper knowledge of the content they are learning because they see the relevance of what they are learning and make more connections to other topics. Likewise, multilanguage learners and students who are neurodiverse benefit from these discourse moves because their voices are heard, and they come to see themselves as doers of mathematics. Understanding types of discourse moves that promote agency and identity is part of the foundation for implementing asset-based instructional routines.

Let's now look at how implementing certain discourse-based routines makes them more or less asset based for both students and teachers (see Table 6.3).

TABLE 6.3   How Discourse-Based Routines Can Create Asset-Based Learning Environments for Both Students and Teachers

ASSET-BASED ENVIRONMENT FOR STUDENTS	ASSET-BASED ENVIRONMENT FOR TEACHERS
Students build confidence in sharing ideas and strategies.	Teachers build their belief in all students as mathematicians.
Students listen to peers and value their contributions.	Teachers position all students as capable.
Students recognize mathematics as a web of ideas where reasoning and sense making are central to learning.	Teachers cultivate a community of learning in which students are engaged, feel capable of learning meaningful content, and use Social Emotional Learning (SEL) to also cultivate the community of learning.
Students think, create, and share ideas.	Teachers intentionally situate the student as the owners of the mathematics.

In an asset-based learning environment classroom, students speak more than the teacher and their reasoning and sense making is valued as students learn new material. Students must feel safe and have a sense of belonging as a foundation for risk taking and sharing ideas in class. Part of the process for creating a safe environment is to make sure we position all students in our classroom as capable and that they feel that their contribution matters when speaking to their teacher and peers. Our efforts to hear our students' ideas are amplified when we have solid instructional routines in place that elicit discourse moves.

ASSET-BASED ROUTINES AND SOCIAL EMOTIONAL LEARNING

Asset-based instructional routines emphasize students' individual and collective strengths. They position students to increase effective collaborative interactions and build relationships. Students listen to, respect, and learn from one another. As such, asset-based instructional routines incorporate NCTM's Effective Teaching Practices while serving two important purposes:

- supporting the learning and teaching of mathematics content and practice; and
- supporting the long-term development of student identity in ways that are consistent with the tenets of SEL.

The asset-based routines in the classroom are not separate from SEL but enhance SEL skills. Let's look at the instructional routines we completed in the Alignment exercise. For each asset-based routine, we will unpack how the routine:

- Facilitates meaningful mathematical discourse, including student-to-student discourse
- Integrates facets of SEL
- Extends and deepens content knowledge
- Recognizes students as the owners of the mathematics they are learning

THINK-PAIR-SHARE (OR THINK-WRITE-PAIR-SHARE)

Think/Write-Pair-Share (TPS) is an asset-based instructional routine that incorporates effective discourse, builds student confidence by asking students to first think independently about a situation, and provides teachers the opportunity to hear student thinking and listen for what students understand. It starts with an open question or prompt given to the class.

How does TPS promote a community of learning? How does this routine support student agency? The prompt should require thinking and explanation (not a specific answer). The initial part of this routine (thinking) promotes a community of learning by ensuring all students have time to generate their own ideas. Writing (or drawing) their ideas supports their thinking process and makes their thinking visible to both them and others. Implementing these two parts (Think/Write) in a TPS is essential in cultivating student's agency and ownership of mathematics. The opportunity to share work with a partner provides students with practice in communication, both speaking and listening. In addition, because students have practiced speaking or showing their work to a peer, they are less anxious and more prepared to share with a larger group. This is especially true for students who would not typically volunteer to share. As always, it is best to know our students and to ask them in advance if they would be willing to share and respect their response if they say they are not.

 Throughout this routine, SEL is developed in several ways. Students develop self-management by learning to manage their thoughts and behaviors by learning to speak and listen to each other. Students can become more self-aware of their thoughts and values and recognize how their behavior influences others. Students grow in their social awareness when they have several experiences collaborating with different classmates and are given structures that position each person as an equal contributor. Students learn to understand the perspectives of others, including those whose backgrounds and experiences may be different than their own. Relationship skills are cultivated through classroom norms that amplify student-to-student interactions from a space of listening with curiosity and wondering (versus one person telling the other person what to do or how to do it). Think about how, over time, responsible decision-making becomes part of a classroom culture as students make caring and constructive choices regarding student-to-student discourse and whole-group discourse interactions. Imagine the change this could make.

> **Tip**
>
> When implementing a Think/Write-Pair-Share experience, select a prompt that allows student choice and encourages thinking versus a prompt that focuses on a specific answer or using a specific pathway to a solution. Be clear about expectations without giving away the mathematics you want them to discuss.

The content knowledge students discuss and engage in effective discourse moves in Think/Write-Pair-Share will vary based on the prompt or question. Several routines incorporate Think/Write-Pair-Share, including Number Talks, the Teaching Cycle from the Comprehensive Mathematics Instruction Framework (CMI) (Hendrickson et al., 2008), and Pause and Reflect. In each of these routines, a prompt is posed (tasks, situations, questions, or context). The posed prompt is intended to promote independent think time, followed by discourse that guides students to the intended learning goals or outcomes.

NUMBER TALKS

Number talks begin with a problem that invites multiple ways of thinking. Although number talks are most used in the elementary grades, they are appropriate across the K–16 spectrum (Rineck, 2020). Then students use mental math to solve the problem (e.g., Humphreys & Parker, 2023; Joswick & Taylor, 2022). Once students solve the problem, they share answers. After the students have determined the solution, the real work of building an understanding of strategies through describing their processes for solving the problem begins. As students explain their strategies, the teacher captures student thinking by writing it down, emphasizing the underlying mathematics rather than correctness. The teacher formulates their questioning and direction of the discussion based on the student's description, placing student thinking at the center as opposed to focusing on the teacher's (or textbook's) way of thinking. Like other TPS routines, teachers invite students to share what they understand without judgment or evaluation and build and comment on one another's thinking.

The asset-based routine of Number Talks addresses important mathematics content while developing mathematics discourse practices and aspects of SEL (Fletcher & Meador, 2022; Joswick & Taylor, 2022). Students build and deepen their content knowledge by thinking of problems more fluently (becoming more flexible, accurate, and efficient). Students extend their ability to engage in discourse by learning how to become very specific in explaining their strategy (either to a peer or to the teacher). Students also have the potential to build all five aspects of the CASEL framework through Number Talks similar to what was described for TPS. For example, students engage in self-management when listening to peers so that they understand the reasoning of their peers in solving problems. Joswick and Taylor (2022) recommended focusing on one or two SEL competencies to develop over time. The collaborative aspects of the routine that relate to social emotional learning also position students to each other as assets in the collective endeavor.

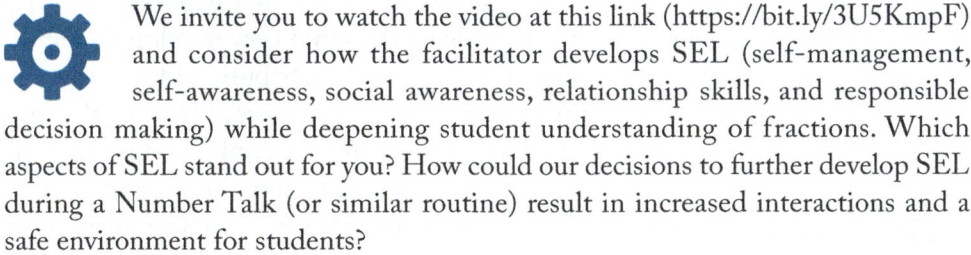 We invite you to watch the video at this link (https://bit.ly/3U5KmpF) and consider how the facilitator develops SEL (self-management, self-awareness, social awareness, relationship skills, and responsible decision making) while deepening student understanding of fractions. Which aspects of SEL stand out for you? How could our decisions to further develop SEL during a Number Talk (or similar routine) result in increased interactions and a safe environment for students?

TEACHING CYCLE

The Teaching Cycle includes three main phases: Launch, Explore, and Discuss (Lemon & Hendrickson, 2023).

- The Launch phase provides context that hooks student engagement in the task and includes relevant information, tools, and expectations for students without giving away the mathematics or directing them to a specific path to solve a problem.

- The Explore phase allows students to work on the worthwhile task independently (Think/Write) before sharing their ideas and strategies with a partner or small group (Pair-share). During this part of the lesson, students make conjectures, critique the reasoning of others, look for structure, use multiple representations to make connections, and revisit/revise their thinking. Students develop ideas, solidify concepts, practice applying their thinking, and build fluency. Meanwhile, during the Explore phase, the teacher monitors students and guides the classroom by posing purposeful questions that assess or advance student thinking. In addition, the teacher identifies students who will share during the whole-group Discuss phase.

- The Discuss phase allows selected students to share their reasoning using their work (whole-group share). Through an orchestrated discussion, the teacher guides the class in producing co-crafted takeaways that align with the lesson's goals.

Tip

Use various methods for selecting students to share in whole group to ensure all students will eventually be chosen and seen as contributors to class discussions.

Tip

Throughout the Teaching Cycle, support all students by providing student choice of how they communicate (writing, drawing, speaking) and how they solve problems such as allowing students to use multiple representations (tables, graphs, equations, pictures, etc.).

 In the Teaching Cycle, students have opportunities to engage in all aspects of SEL (self-management, self-awareness, social awareness, relationship skills, and responsible decision-making) over time. Students learn to share strategies with peers without being prompted by the teacher. Thus, they learn to recognize when other team members have had time to think through the problem on their own before they begin sharing their work. It also means that they develop respect for their peers by both contributing ideas and being open to the ideas of others. Through this discussion and ongoing interactions, students learn that it is ok to adapt their thinking and adjust their work to make it clearer. These adaptations occur because students gain clarity of their own ideas and gain new insights from what they learn from their partner or small group.

PAUSE AND REFLECT

In Pause and Reflect, teachers provide students with a prompt during the lesson that causes reflection on the mathematics they are learning. After reflecting, students summarize key insights, strategic approaches, and possible questions about the mathematics of the task with their peers. This routine allows teachers to have students pause from their work, step back, and pay attention to how they are attending to the task and how it has advanced their thinking. Below is an example of how Ms. Sharp has used this routine to make her sixth-grade students more aware of their strategies to match equivalent expressions.

Ms. Sharp has given her students a task that has them match equivalent expressions. After working independently and in small groups for a while, Ms. Sharp has noticed students using various strategies. She has the class pause their work and asks them to discuss in their partner/small groups what methods or strategies they use to determine whether two expressions are equivalent. Students pause for a moment and then verbally share what they are doing. Ms. Sharp listens to the discussion to informally assess student understanding and select some people to share their thinking. After two minutes, she brings the class back together and has the following discussion:

Ms. Sharp: "Jayden, will you please share what your group did to match equivalent expressions?"

Jayden: "We took the variable x and turned it into a 2, so when the problem said $x + x + x$, we said it was $2 + 2 + 2$, or 6. We then found another expression, $3x$, and when we used two again for x, it was also 6."

Ms. Sharp: "Would it always work to pick a number and see if both expressions have the same answer to show that they are equivalent?"

Jayden: "Not always. Using the same method, we also got 6 for the expression $2(x + 1)$. We felt this wasn't equivalent, so we tried a different number to show they were not the same. My partner said we could use what we learned in elementary school. If we know that $2 + 2 + 2$ is 3 groups of 2, or 3 times 2, then $x + x + x$ is 3 groups of x, or $3x$. This makes sense to me."

Ms. Sharp: "Can someone else restate what Jayden is saying about their strategy?"

Bella: "When your expression has multiplication, you can use what we learned in elementary school. 3 times 2 means 3 groups of 2, or $2 + 2 + 2$. We used the same strategy for $2(x + 1)$ by saying 2 groups of $(x + 1)$, or $(x + 1) + (x + 1)$. This gave us $2x + 2$."

Jayden: "Yes, when we see multiplication, we can think of groups and how many in each group. $5y$ would be five groups of y, or $y + y + y + y + y$."

Ethan: "I think Jayden also used plugging in numbers as a strategy. If the expressions are equal, they should have the same value. If they don't, they don't match. If they do, then we have to show why or find a different number to show they are not equal to each other.

Ms. Sharp waits, then nods and says: "Alex, would you like to share what your group discussed using the expression $2(x + 1)$?"

Alex:	"We used the distributive for $2(x + 1)$ and got $2x + 2$.
Ms. Sharp:	"Can you please expand your explanation of how you used the distributive property to get $2x + 2$?"
Alex:	"Yes. We multiplied 2 times x to get the $2x$, then we multiplied 2 times 1 and got 2. So our solution was $2x + 2$."
Ms. Sharp:	"Thank you all for sharing and listening to each other. We just heard a few strategies to assist us in finding matches for equivalent expressions. Please continue your work for the next ten minutes. Pay attention to how you determine if two expressions are equivalent. We shared a few methods, but you may find you are using other strategies or properties. You are working hard, and I appreciate it!"

As seen in the example, students first pause and think about what they are doing mathematically. Sharing with a partner allows the teacher to assess what students understand and for students to verbalize their thoughts. The short, whole-group discussion allows students to reflect on how they are identifying equivalent expressions. Like TPS, this routine builds a community of learning in which there is an opportunity for students to own the mathematics. Ms. Sharp did not tell students how to find equivalent expressions but asked students what they were doing and had them justify their thinking. In addition, students who were stuck or only using one method may now have different ways of thinking about creating or identifying equivalent expressions.

■ ■ ■ Try This

Now is a great time to review the Alignment exercise from the beginning of the chapter. How does effective implementation of the instructional routines we have discussed cultivate a student's positive mathematics identity? What new insights about these routines are needed to effectively implement student discourse? You may notice that the routines by themselves may be designed to be asset based but that, ultimately, it is how we implement the routines that determine where they fall along the continuum. I (Joleigh) have had my share of instances when my efforts to have students think, reason, and communicate have fallen flat. It takes time, practice, and reflection to make instructional routines fall further along the asset-based continuum. Review the *Features that cultivate a student's positive mathematics identity* in the Study Guide and add new ideas to your list that will enhance your own implementation of each routine. ■

online resources ☞ The online study guide for this book is available for download at https://qrs.ly/xyfid21

TWO-MINUTE TALKS

A Two-Minute Talk is an asset-based instructional routine that encourages students to communicate information to a partner using speaking and listening skills. Students are paired up, and then each person spends one minute sharing their knowledge about a given concept or topic (Seda & Brown, 2021). Through this routine, students access their background knowledge as they share their understanding about a topic with their partner. Likewise, they build and deepen their knowledge about the topic by listening to the contributions of their partner. This student-to-student interaction cultivates a community of learning and positions students as the owners of the mathematics they are sharing. This routine works best after students have spent time on a particular concept. As students reflect, share, and listen, they self-assess their understanding of what they know, what they are still learning, and where they may need additional support. This short routine also allows the teacher to listen and informally assess individual and collective strengths students have about the concept and notice areas that may need attention.

Using Two-Minute Talks allows students to examine and extend their knowledge and develop social emotional (academic) learning. Students learn to self-manage their thoughts and behaviors when they are listening and when they are speaking with their partner. Learning to communicate academic ideas includes speaking confidently but not in an overbearing way and listening from a space of curiosity while critiquing the reasoning of others. This work means that students recognize their peers as having mathematical ideas worth listening to and that they can contribute to the knowledge of others. Table 6.4 highlights aspects of SEL and how this routine supports student development.

TABLE 6.4 How Two-Minute Talks Support SEL Development

SOCIAL EMOTIONAL LEARNING	DEVELOPMENT OF SEL
Self-manage	Students learn appropriate behaviors when sharing and interacting with peers
Social awareness	Students learn to value and appreciate the contributions of peers, including when those contributions bring different perspectives
Relationship skills	Student interactions cultivate supportive relationships

TURN AND TALK

Turn and Talk is a short instructional routine in which students turn and talk to their partner about a prompt presented by the teacher connected to a current discussion occurring in the classroom. This routine can fall in many places along the deficit-to-asset continuum, depending on the implementation. The more clarity of expectations, the more open the prompt, and the more direction provided, the more this routine is asset based. Additional considerations that make a huge difference include using classroom norms (Chapter 3) and prompts that include explanations from students.

Here are some examples of actions that *reduce* a safe learning environment or have this routine fall more on the deficit side of the continuum:

- failing to provide students appropriate time to think,

- not setting up the routine so that each partner has equal speaking/listening time, and

- not considering language support to remove communication barriers (Walter, 2018).

Turn and Talk can be a powerful routine for students when given concise, clear expectations and time to think on their own before engaging in discourse with a partner. This routine can quickly become deficit based if not implemented thoughtfully. For example, if the directions are unclear, too wordy, or expectations are not stated, students may feel less confident and defer to their peers to do the talking. This is particularly true for students who have been historically marginalized and have heard messages, explicit or implicit, that mathematics is not for them.

Be sure to include who shares first in the directions, so students who process quickly and are often eager to share benefit from this routine by learning to listen to the ideas of others and recognize that different perspectives can increase our own understanding. We want to make sure it isn't always partner A who begins. For example, the teacher might say, "Please consider this expression on the board and be ready to explain its meaning to your partner. I am going to give you 30 seconds to think about this. (30 seconds passes.) Okay, if you are the partner sitting closest to the window, you will share first. You have 30 seconds to present your thoughts and then your partner will have an opportunity. Go!"

> **Tip**
>
> Develop a variety of strategies to determine who goes first. Examples include the person with longest (or shortest) hair, sitting closest (or furthest) to the door (or window or closet or board), or by placing colors or shapes on desks so that the person with the blue shape, or the rectangles, go first. You get the idea—mix it up!

Let's look at an example of Turn and Talk in Ms. Love's class. Can you identify the ways she uses information to build student learning?

Students in a seventh-grade math class have been working on developing their understanding of probability. In pairs, students designed a survey and collected data in their class about a topic of their choice. Each group obtained a data set of 30 responses from their classmates and then projected what the results would look like if they asked all seventh graders in their school (250 students). After students summarized their findings and made explicit connections to proportional reasoning and probability, the teacher, Ms. Love, posed the following question:

- "How would you predict the results of your survey if we wanted to predict the results of a larger group, such as the entire school, the entire community, or the entire state? Is this possible with the data you collected?"

- "Turn and Talk to your partner. Partner A is the person sitting closest to the door. Partner A will explain their process for making predictions of a larger group using your data. Partner B will agree/disagree and say why, then will explain whether the prediction would make sense to a broader audience. Partner A then agrees/disagrees and shares why. The prompt, timing and sentence stems for each partner are on the board. Partner A may begin their thoughts and explanation now."

  Partner A: 90 seconds:

  Explain your process for making predictions of a larger group using your data.

  Sentence stem: "My process for making predictions of what our data would look like for a larger group would be to _____."

Partner B: 90 seconds

Agree/disagree and say why, then explain whether the prediction would make sense to a broader audience.

Sentence stem: "I agree/disagree with your process because _____.

Our data would/would not make sense for a larger audience because _____."

Partner A: 30 seconds:

Agree/disagree with partner B's statements and explain their thinking.

Sentence stem: "Thank you for sharing your thoughts. I agree/disagree with what you said because _____."

 As students discuss, the teacher listens to the ideas being shared around the room. After four minutes, Ms. Love identified key ideas to surface in a whole-group discussion.

Ms. Love: "Rochelle, can you please share what your team discussed about the process for predicting data for a larger audience?"

Rochelle: "We said you could use the proportion strategy Micah used yesterday. The first ratio would be our data from our class (with 30 in the bottom) and the other ratio would have an x on top, and the number of people in the data on the bottom. For us, it would be 7 people say soccer is their favorite sport out of 30 of us for the first ratio, and then we would write equals x people say soccer is their favorite sport out of 30,000 people. We think there are 30,000 people in our community so 7,000 people say soccer is their favorite sport."

Ms. Love: "Thank you, Rochelle. Questions for Rochelle? Did anyone use a different strategy?"

Daisy: "We created a graph. The line extends so that you can predict for any size group."

Ms. Love: "Daisy, can you use Rochelle and Micah's data about soccer and draw the graph on the board to show us what you are talking about?"

Daisy goes to the board and creates a graph, starting at $(0, 0)$ and draws a line that goes through $(7, 30)$. She writes "total people" on the y-axis and "soccer is my favorite" on the x-axis. She then continues the line and places $(70, 300)$ and $(700, 3000)$ on the graph and says, "We can pick any points on this line and they would fit in our table as a prediction."

Ms. Love: "Rochelle and Daisy, thank you for showing us two different strategies for making predictions about our data for larger audiences. Everyone, take a moment to write down how you can use different representations to make predictions. You may have also used a different representation than what was shared so be sure to include it. We will discuss these in our takeaways in a bit. The teacher pauses so students can record their thinking and then continues, "Jayden, can you please share what you and your partner discussed about making predictions to a larger audience?"

Jayden: "James and I collected data about how much time we spend on homework each week. We said that our data is probably good for seventh graders at our school, and maybe seventh graders at other schools, but would not make sense for adults who don't do homework."

The teacher pauses to give other students an opportunity to consider what was just shared and to provide space for student-to-student discourse in a whole-group setting.

Tanesha: "Our data was like that. Our survey asked, 'how much time do you spend on your phone?'"

Micah: "Yeah, now that I think about it, I am not sure that our class data would be a good prediction for how many people say soccer is their favorite sport. I think it would be a lot higher."

Rochelle: "I think it would be lower. No one in my family even likes soccer."

Dwayne: "Our survey showed that football was most people's favorite sport."

Ms. Love: "Thank you all for sharing your thoughts on this. I wonder what makes data reliable when we go to make predictions that go beyond the data we collect? We will pick up this discussion in tomorrow's lesson. For today, let's summarize different strategies we can use to make predictions based on the data we collect."

How do the instructional routines of Two-Minute Talks and Turn and Talk cultivate a student's positive mathematics identity? What new insights do you have about these routines to more effectively implement student discourse? Review the *Features that cultivate a student's positive mathematics identity* in the Study Guide and add new ideas to your list. ■

OBSERVATION ROUTINES

Observation-focused instructional routines are perfect for creating a safe environment where students and teachers have opportunities to notice, wonder, and analyze mathematical characteristics. Because these are observations, our role as the teacher is to focus on what the students are connecting to without judgment. As students observe through noticing, analyzing, or comparing (seeing both similarities and differences), they generate lists or recognize features based on what they observe. Since student observations drive the work, students are the owners of what is shared, and they gain insights from observations from their peers. The emphasis in these routines is for students to make sense of and explain their observations. As educators we do and should anticipate student responses; however, the focus is always on what the student brings to the discussion and not on whether they made "the right observation" or selected the "correct choice." Currently, the two most popular asset-based math instructional routines that use this strategy are *Which One Doesn't Belong?* and *Notice and Wonder*. Let's look at these two routines more closely.

WHICH ONE DOESN'T BELONG?

Which One Doesn't Belong? (or Which One Is Most Unique?) highlights student thinking and reasoning as opposed to answers that are right versus wrong. Students reason about each option and then use mathematics to identify and explain how their choice is most different or unique (Danielson, 2016).

Similar to Which One Doesn't Belong, Notice and Wonder removes the pressure of having the "right answer" and allows multiple entry points for students to engage in a mathematical situation. According to NCTM Past President Trena Wilkerson, "When we share a picture, a numberless problem situation, a graph, or even a pair of equations or number sentences and then ask students what they notice and what they wonder, we engage them as active thinkers and doers of mathematics.

We position them as authorities in the classroom, valuing their voice and ideas" (Wilkerson, 2021, para. 1).

Let's look at an example of *Notice and Wonder* from Mr. Wright's eighth-grade math class:

Mr. Wright displays the following graph on the board.

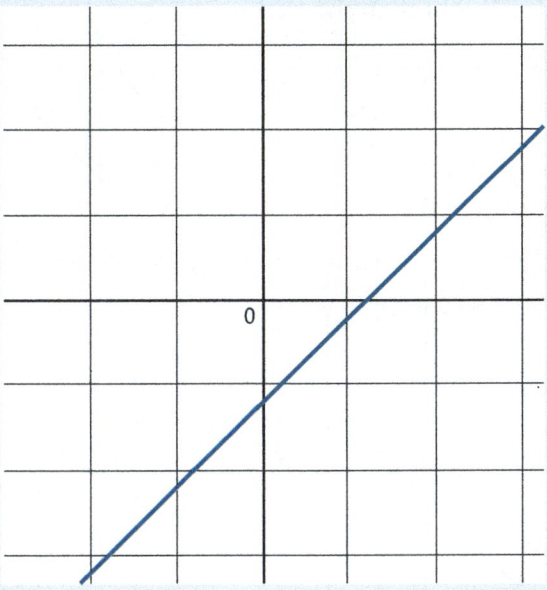

Mr. Wright: "We are going to engage in a Notice and Wonder routine to start our day. What do you notice?"

Dave: "It is a linear function"

Aly: "The slope is positive"

Shawn: "The graph is going up"

Abdul: "The *y*-intercept is negative"

Juan: "There are no numbers on the graph"

Mr. Wright captures student comments on the board. After a moment, Mr. Wright continues by asking, "Thank you for sharing what you have noticed. Using the same graph, what do you wonder?"

Abdul: "I wonder if there is another *x*-intercept"

Veronica: "I wonder what the equation is"

Tate: "I wonder what this has to do with me"

Alisha:	"I wonder what the graph represents"
Mr. Wright:	"I appreciate your comments, including both what you noticed and what you are wondering about. I am going to reveal more information about this graph and would like you to find the equation of this line. After you write your equation, create a story of what this graph could represent and label your axes based on your story." Mr. Wright then shows this:

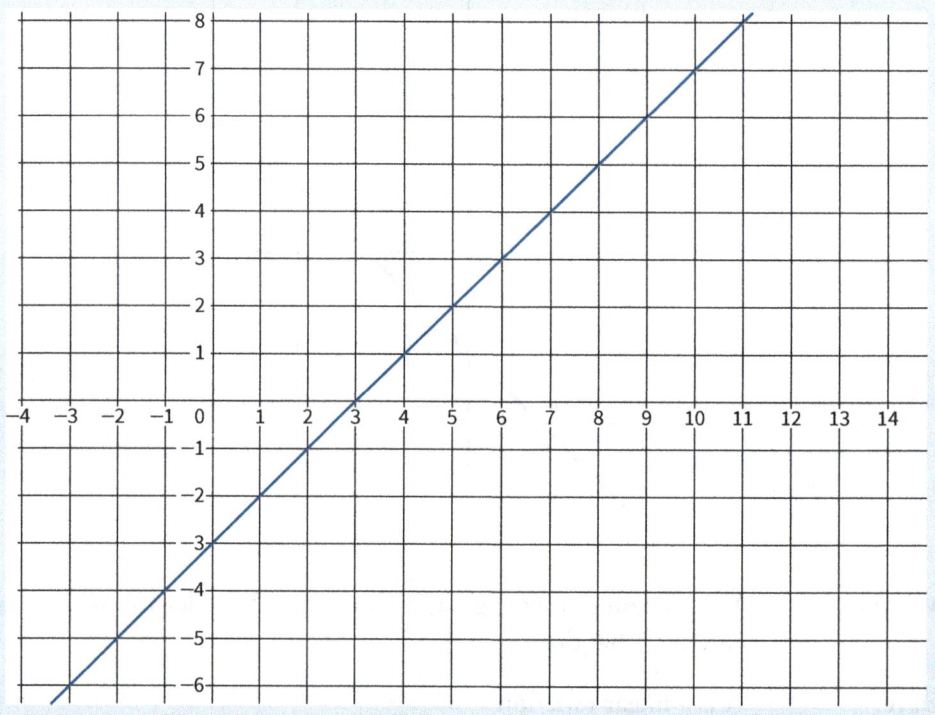

Mr. Wright:	"Use what we noticed to see if your equation, story, and our noticings all line up. Be prepared to share with a partner."

This short routine is a catalyst for student thinking and curiosity. In Mr. Wright's class, students have accessed their background knowledge to notice and wonder about the given graph with the initial function and can use this information to further make sense of the complete graph as they tell the story. This example shows an implementation that is very asset based as it positions students as valuable contributors, promotes a safe environment, and builds a community of learning. Both Notice and Wonder and Which One Doesn't Belong fall far along the asset-based side of the continuum when the routines are implemented appropriately. They are less asset based (and can even become deficit based) if we do things like listen for the "best idea" or dismiss an answer we do not anticipate. This has the opposite impact of what we intend and can in fact move us away from a safe environment where students either compete to share the "best" answer or stop sharing ideas in general.

■ ■ ■ Try This

Try cultivating curious and creative problem solvers by using the Notice and Wonder routine to engage students. For example, you might ask students to engage in a quiet write: "Write down everything you know about a parade?" Give students time to think (and to write their thoughts down). After about a minute, ask the class to share with a partner. Let students share and then ask them to popcorn their answers (shout out a word or phrase one at a time) as you point to different partner groups. You might get responses like:

- music,
- floats,
- cars,
- people on the floats, and
- people on the side who watch and cheer.

The intention is to have students think and share but also to build a community that shares ideas. Acknowledge all responses. This can be as simple as a head nod after each response and then saying "thank you for sharing" to the whole class once answers have been shared. Next, play a short video of the Macy's

Thanksgiving Day parade (you can find these on YouTube) and ask students to watch and take notes about what mathematics they notice and wonder about in the parade. After two minutes, stop the video and let students finish writing their notice and wonder. Then direct them to share with a partner. After a couple of minutes, call on a few students to share their wonders with the class.

Their responses might include:

- "How many people does it take to hold the balloons?"
- "How much helium is used in a balloon?"
- "How long does it take to walk the parade?"
- "How long does it take to fill up a balloon?"
- "How much material is used to make a balloon?"

Use the responses to direct students to determine a process for finding the amount of material it would take to create a particular balloon. ■

How do What Do You Notice? What Do You Wonder? and Which One Doesn't Belong? cultivate a student's positive mathematics identity? What new insights do you have about these routines to more effectively implement student discourse? Review the *Features that cultivate a student's positive mathematics identity* in the Study Guide and add new ideas to your list. ■

MATHEMATICS LANGUAGE ROUTINES (MLRs)

Mathematics language routine (MLR) support students in learning language, as well as mathematics content and practices. A research group out of Stanford University created MLRs that were built using four asset-based design principles that see students as capable of using, implementing, discussing, and analyzing sophisticated mathematical concepts and terms (Zwiers et al., 2017). The four design principles foundational to the MLRs include:

- support sense-making,

- optimize output,

- cultivate conversation,

- maximize meta-awareness.

From there the following eight MLRs were created:

- MLR 1: Stronger and Clearer Each Time

- MLR 2: Collect and Display

- MLR 3: Clarify, Critique, and Correct

- MLR 4: Information Gap

- MLR 5: Co-Craft Questions and Problems

- MLR 6: Three Reads

- MLR 7: Compare and Connect

- MLR 8: Discussion Supports

Each MLR is intended to develop language and, when implemented as intended, increase student identity. In this chapter, we will focus on Routine 7: Compare and Connect and on Routine 8: Discussion Supports. These two routines can be incorporated into your current instruction and have strong impact with little effort. For more information about all of the MLRs, as well as their development, visit http://ell.stanford.edu/content/mathematics-resources-additional-resources.

MLR 7: Compare and Connect

A well-designed mathematical task will include multiple entry points with various solution paths. Thus, students will naturally have different perspectives as they engage in problem solving. These different perspectives generate work from students that beautifully creates opportunities for comparing and connecting mathematical ideas. This routine can deepen student content knowledge in many ways while providing opportunities for discourse. Asking a class to compare two pieces of work increases students' metacognition as they analyze similarities and differences. For example, students become more flexible with manipulating expressions if they spend time comparing expressions that are equivalent but written in different ways. Another example is when students compare and connect different functions to gain a deeper understanding of their characteristics.

As students learn to compare and connect, they can begin to use this routine as a strategy for problem solving. For example, as students become confident in making connections between representations of a function, representations become a tool to solve other problems, hence increasing their ownership of the mathematics they are learning. They see themselves and their peers as authors of mathematics because it is the work generated by the class that they are using to make comparisons and connections.

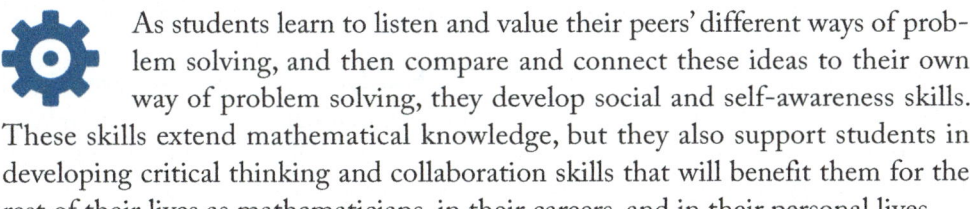 As students learn to listen and value their peers' different ways of problem solving, and then compare and connect these ideas to their own way of problem solving, they develop social and self-awareness skills. These skills extend mathematical knowledge, but they also support students in developing critical thinking and collaboration skills that will benefit them for the rest of their lives as mathematicians, in their careers, and in their personal lives.

MLR 8: Discussion Supports

Discussion Supports are used with other discourse-oriented routines to scaffold student-to-student interactions, build a community of learning, and overall incentivize meaningful classroom conversations. Examples of discussion supports include implementing sentence stems, referring to strategies using the name of the student who first presented work using the strategy, and intentionally building student-to-student questioning and discourse in the classroom.

Sentence stems are widely known to be effective in supporting multi-language learners to engage in mathematical discussions by removing language barriers and amplifying student thinking by including explanations or justifications. In addition to multilanguage learners, sentence stems help students who identify as neurodiverse and benefit from the structure. Sentence stems are also good for the entire class as they cultivate a safe environment by setting up the structure that both students share using the same or a similar structure. Sentence stems can also reinforce norms in the classroom. For example, the sentence stem "I think _____ because _____" has a different feel than if one student says to their partner, "The answer is _____ because _____."

Discussion supports can also be used to build appreciation of peer contributions. When students share their work, the teacher scaffolds student interactions such that students begin using these strategies themselves to prompt each other to engage more deeply in discussions.

■ ■ ■ Try This

How does each MLR discussed in this chapter (Compare and Connect and Discussion Supports) cultivate a student's positive mathematics identity? What new insights do you have about these routines to more effectively implement student discourse? Review the *Features that cultivate a student's positive mathematics identity* in the Study Guide and add new ideas to your list. ■

THE CASE OF "I DO, WE DO, YOU DO"

The instructional routine of I Do, We Do, You Do is a gradual release model, where the teacher holds the knowledge and then gradually releases the control to the students. When implemented from this point of view, the routine typically begins with the teacher telling students information (usually going through steps in a procedure), then has students and the teacher solving similar problems together following the same process, and then eventually releasing students to independently solve more problems using the same process and steps. The implementation of the routine looks like this:

- I Do: The teacher shares their process for solving problems, sending the message "this is how I think."

- We Do: The teacher supports students in working similar problems, sending the message "we all think like me" and "we solve all problems like this using this process."

- You Do: The teacher releases students to independently solve problems using the same process, sending the message "think like me" (Ramirez, 2021, ASSM).

Where do you think this routine falls on the deficit-to-asset continuum?

When we implement this routine so that the focus is on following steps to solve a specific problem as opposed to developing problem-solving skills, it does not position students as the owners of the mathematics. If we use this as our go-to routine for initial learning of content, then we miss opportunities to access students' background knowledge and for them to share what they know. When we send the message that math is a set of steps to be followed, then we create a deficit-based understanding of what it means to know and do mathematics.

However, like all routines in this chapter, *how* we implement a routine has greater impact on where it falls along the continuum than the routine itself. This routine falls along the asset-based continuum when used to introduce notation or a mathematical convention based on student contributions. In this situation, a teacher may use student thinking or work to introduce an explicit strategy, vocabulary, notation, or convention. For example, when a student identifies potential solutions in a coordinate plane for a linear inequality, it is appropriate for the teacher to explicitly explain how we show all potential solutions by shading the area where the solutions exist. In this situation, the teacher explains the convention we use as mathematicians and supports students in using this convention moving forward. The important takeaway for this routine is that it is best when used in conjunction with other asset-based routines and is not the primary form of instruction.

■ ■ ■ Digging Deeper
"Exceptional Students"

Let's dig deeper into why discourse and building a positive mathematics identity is important for students who identify as having a propensity for mathematics (often being placed in Honors-type mathematics courses).

Students in Honors courses and/or those identified as having a propensity in math have been trained over the years that they are good at math because they are fast at solving problems. Although specialists will say students need curiosity, creativity, and problem solving, most Honors classrooms move more quickly through curriculum, not necessarily more deeply into understanding the content. The result: Students in Honors courses often describe math as a class about answer getting. They are also prone to imposter syndrome, worrying that if they do not grasp the information as quickly as their peers, they aren't really as good at math as people expect them to be. Fast is equated with "smart." Not getting it fast often translates to a negative mathematics identity.

In my role as a state specialist working with districts who were implementing Learning Intentions and Success Criteria, I interviewed more than 200 secondary mathematics students enrolled in Honors classes, asking them the following three questions:

1. What are you learning?

2. Why are you learning this concept?

3. How will you know you have learned it?

More than 80% of students were able to answer *"What* are you learning?" but often struggled with responding to the other two questions. Students who described math class as answer getting said they were learning the concept so they could take more advanced math and that they would know they have learned it based on their homework, quiz, or test score.

Students in discourse-rich classes were more likely to provide explanations about why the concept was important and to provide success criteria or an explanation for how they would know they had learned the mathematical concept.

We want all of our students to realize the joy, beauty, and wonder of mathematics. What is the message we are sending to students enrolled in Honors classes if most do not know why they are learning math concepts? Now more than ever, we must work to provide students with relevant mathematics so that students recognize how the concepts relate to their world and see themselves as *doers* of math. ∎

Reflect, Apply, Transform

Consider the questions posed at the beginning of the chapter:

1. What instructional routines contribute to a student's positive mathematics identity?

2. How do I implement asset-based instructional routines that cultivate a positive mathematics identity and support deep mathematics learning?

Using your answers to these questions, select one to three instructional routines that you would like to take to the next level in your classroom, moving this routine further along the asset-based continuum. You may wish to review your responses to these routines in the *Features that cultivate a student's positive mathematics identity* in the Study Guide. Next, plan discourse that highlights what students know and positions students as *doers* of mathematics using the Implementation of *Asset-Based Routines Template* in the Study Guide. This template can be completed for a lesson or a more broad concept. An example of the Template completed is shown in Table 6.5.

online resources ⯈ The online study guide for this book is available for download at https://qrs.ly/xyfid21

TABLE 6.5   Completed Example of the Implementation of Asset-Based Routines Template

Part A: Unpack the Concept	
Concept students are learning	Quadratic Rates of Change
Where are students in the progression of learning (developing conceptual understanding, solidifying their understanding, or are they practicing and building fluency) with the concept?	New Concept—developing conceptual understanding
How will you provide access to the content so that all students are positioned as capable?	Access background knowledge Allow student choice for representations Provide individual think time before sharing with a partner
Part B: Instructional Routine (or Routines)	
Which asset-based routine (or routines) will you use? *Use Part A to make decisions when selecting routines.*	Think/Write-Pair-Share Discussion Supports (Sentence Frames) Turn and Talk
How will you use this routine to focus on what students know (as opposed to what they do not know)?	Students will be given a pattern and asked to use representations of their choice to make sense of the pattern. I will look and listen for how students describe how they see the pattern grow. I will ask questions for clarification, not to redirect. I will select a few students who have used different representations to share their ideas with the whole class. Students may use knowledge of linear/exponential functions to compare rates of change.
How will you position all students as capable?	Students will be given independent think time and choice in how they make sense of the pattern before sharing with peers.

(Continued)

TABLE 6.5 (*Continued*)

Plan in Action:

1. Students are provided a pattern and asked to make sense of how they see the pattern grow using multiple representations.

2. Students are provided five minutes of independent think time to make their thinking visible and to make connections between representations they have created.

3. Students share their work with a partner and compare how they thought about how the pattern grows using the following sentence frame: "I used a _____ (name of representation)_____ and noticed that the pattern grows like _____." Students continue working together to generate other representations and to see if they notice other characteristics of how this pattern grows.

4. I (the teacher) will select students to share their observations with the class (at least two students who have used different representations).

5. After two students share, I will ask the class to turn and talk with their partner and share what is similar and what is different about the students who shared. Sentence frame (partner A first, then B): "I noticed that they are similar because _____" . . . after both share, then Sentence frame (Partner B first, then A): "I noticed they are different because _____."

Questions to consider—does my plan answer these questions? What other additions might I make if something is missing?

*   What instructional routines contribute to a student's positive mathematics identity?

*   How do asset-based instructional routines both cultivate a positive mathematics identity and support deep learning of mathematics?

CHAPTER 7

Lesson Routines That Become Practices

The previous two chapters have focused on routines that operate at a smaller scale as we considered structural routines and how routines can help us know what students know. But as teachers, we know that lessons aren't just collections of routines. For example, I don't just decide to use a think-pair-share or a turn-and-talk routine at random in a lesson. Rather, I'm thinking about the storyline of the lesson and where I expect there to be deep and thorny mathematical ideas that would benefit from student conversations. I'm thinking about where there are plot twists in the mathematical storyline—unexpected and surprising mathematical ideas where I'd like students to take a moment to catch their breath and wonder about what just happened and what might happen next. When we build routines thoughtfully into lessons, it makes us better able to leverage student assets at exactly the right times in lessons so that students see themselves not just as occasional contributors to math class but also as authors and discoverers of powerful math ideas. Thinking at the lesson level also increases the power of our routines—when used together, they have the potential to elicit stronger student contributions than they might when used individually.

In this chapter, we'll explore how routines can become practices that scaffold ongoing lesson-level work and consider how those sets of routines coalesce to support you and your students doing important mathematical work together. The Alignment exercise that we'll begin with here asks you to think about how important mathematical ideas are built with your students in the course of a lesson.

> *[Routines like] The Five Practices are* doable *and something teachers could continue to get better at doing over time.*
>
> *(Smith, Steele, & Sherin, 2020, p. xxv)*

Questions to Consider

1. In what ways do our routines for planning and teaching mathematics reflect asset-based perspectives?

2. How do our routines resonate or conflict with aspects of asset-based language?

3. How can routines foster Social Emotional Learning (SEL) competencies and support access for all learners?

4. What aspects of our routines are ripe for change?

ALIGNMENT EXERCISE: THINKING THROUGH A FAVORITE LESSON

As teachers, we frequently have favorite lessons that we look forward to teaching every year. There are a million different reasons why we might look forward to that lesson—it might be math content that we love, it might be something about the task we designed for the lesson that we really enjoy, or it might be the joy we see as students confront a new mathematical conception that challenges a preconception. Take a moment to think about one of your favorite lessons that fits this mold. We'll ask you to do some reflective analysis of that lesson in this Alignment exercise.

One lesson that I (Mike) remember fondly in this space was teaching box-and-whisker plots in middle school. Early in my career, I realized that students struggled to grasp the concept that each section of the box-and-whisker plot represented the same number of data points. To address this, I designed an activity in which students themselves were the data points and we built a human box-and-whisker plot on the sidewalk in front of my classroom. Students held labeled meter sticks that represented the quartile boundaries. It was a lesson I was proud of and looked forward to every year. But as I had continued to implement more discussion-focused routines, something just didn't sit well with me about this lesson. (I'll share my reflections on that point at the end.)

■ ■ ■ Try This

Here's what we'd like you to think about with your favorite lessons. Walk through the lesson in your mind and think about the storyline. What's the beginning, middle, and end of the lesson? You might even draw out a storyboard for the lesson that illustrates what happens in each section. What's the key idea at each point? What are you doing, and what are your students doing? List for yourself the routines that are in place for the lesson.

Now, what do you notice about these routines? Where are the opportunities to elicit student thinking and to surface and use students' mathematical assets? Where is the lesson more driven by you as the teacher? ■

One thing that I noticed when reflecting on my human box-and-whisker plot lesson was that although it may have illustrated a key point that I wanted students to understand, very few opportunities existed to leverage student assets. Students followed the procedure I had laid out to create the plot on the sidewalk, dutifully arranging themselves as a data set. But students had few, if any, moments in the lesson until the very end to share their thinking or make observations grounded in their experiences. Student dialogue largely consisted of questions about what to do and why they were taking this particular step in the procedure that I had laid out. At the end when the plot was created, I had envisioned this moment as the place for students to make observations. But because they served as the data points, they literally could not see the big picture—only I, as the teacher standing back 30 feet away from the sidewalk's edge, could see the entirety of what was happening. I found myself subtly prompting the observations I wanted students to make rather than allowing them to genuinely leverage their assets. As I had been opening up the discourse in other lessons in my class and providing more opportunities for students to share the assets they brought to the math classroom, I realized that despite my love of the lesson, it was not sitting well with me because the more directive routines I was using were not consistent with an asset-based approach.

In this chapter, we invite you to think about how your routines fit together in the context of a complete lesson. We'll draw on the work we've considered about the nature of our routines in the previous two chapters and consider how we can use asset-based routines together to frame a storyline for a lesson.

WHAT'S THE ROLE OF OUR PLANNING AND TEACHING IN FOSTERING ASSET-BASED ROUTINES?

As we've seen so far, individual routines can be used in ways that fall in different places on the deficit-to-asset continuum. The way we use individual routines is undoubtedly important. Our routines can be even more powerful when we plan to combine them thoughtfully across the course of a lesson. The goal of our work together in this chapter is to illustrate how combining routines together can open up new opportunities to center student assets in our math teaching. We'll also consider how sets of routines together can create opportunities to address and develop aspects of Social Emotional Learning (SEL) as described by the Collaborative for Academic, Social, and Emotional Learning (CASEL) Framework (2024).

To do this work, we turn to the *5 Practices for Orchestrating Productive Discussions* (Smith & Stein, 2018). This framework is a collection of routines that evolved from work Peg Smith and Mary Kay Stein did teaching teachers to use inquiry-oriented mathematics teaching and supporting them as they did that work. Specifically, they noticed teachers (including themselves) using a set of routines in predictable ways to orchestrate rich math discussions. They noticed patterns in how the teachers thought through what students might say or do, the questions they planned to assess and advance the nature of student thinking, how they organized activity with small- and whole-group discussion, and who was selected to share which math ideas publicly and in what order. As we visit the routines that make up the 5 Practices, we encourage you to consider how you think about combining your routines across the course of a lesson to create opportunities for students to share and leverage their mathematical assets.

DESIGNING LESSONS THAT LEVERAGE ASSET-FOCUSED ROUTINES: THE 5 PRACTICES

A central assumption in the 5 Practices work is a particular flow to a lesson (see Figure 7.1). This section provides an overview of the 5 Practices along with connections to asset-based perspectives, Universal Design for Learning (UDL), and SEL. If you are familiar with this work, you may wish to skim this section or move directly to the next section, Example of The 5 Practices in Practice: Scott's Workout. The 5 Practices reflect the Teaching Cycle described in Chapter 6, going into greater depth about the structure of the Launch, Explore, and Discuss phases of the lesson using specific teacher moves.

FIGURE 7.1   5 Practices for Orchestrating Productive Discussions

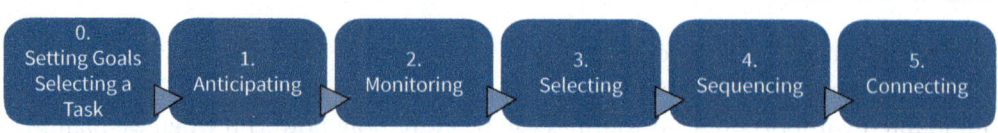

SOURCE: Adapted from Smith and Stein (2018).

0. SETTING GOALS/SELECTING A TASK

The first part of this flow begins before we are in class and includes establishing the mathematical goals and selecting the worthwhile task that aligns with the content we intend to teach. Using our goal(s) and selected task assists us in planning for how the story of the lesson will unfold.

> **Tip**
>
> Select tasks with multiple entry points/solution paths to empower students to choose strategies and use their funds of knowledge. Students will then be supported in becoming confident knowers and doers of mathematics.

1. ANTICIPATING

Once the task is selected and the goals are identified, teachers anticipate various ways a student may think about a problem or situation. Anticipation of student work includes incorporating multiple teacher-focused routines such as considering different strategies students may choose, identifying areas where students may have misconceptions, and recognizing what's valid or correct in student thinking that may otherwise not be an accurate solution. Wondering what students will say during a lesson can be stressful! Careful and thorough anticipation reduces our anxiety during the monitoring phase by shifting the focus from making sense of the student's strategy (already anticipated!) to listening to the student's thinking. Educators who use this practice often share that anticipating has improved their teaching by increasing flexibility of their own thinking and broadening their understanding of content. In addition, this practice alone encourages us as educators to think of how to provide students choice.

For students learning to speak a new language and for students who are neurodiverse, this opens up a world of using drawings and other use of action and expression. Although choice is good for every student, it is especially good for students who otherwise may have difficulty sharing their understanding when confined to only one method. A lack of anticipation often leads to directing students to a specific strategy or procedure the teacher has in mind rather than building from students' thought processes. Anticipating is a critical step to help us determine how to use student thinking to move from one mathematical idea to the next as the lesson unfolds.

> **Tip**
>
> Consider various possibilities/ strategies to solve the problem. This approach supports listening to your students' ways of thinking and then seeing themselves as mathematicians, and it increases your comfort level when seeing different student work.

After anticipating, you'll get students started on the work of the lesson by launching the task.

The launch is short, but this routine includes sharing the context with students to motivate engagement and describing expectations without stealing the joy or giving away the mathematics. Students then engage with the task independently, exploring different solution paths to make sense of the problem on their own before sharing their ideas and strategies in a small-group discussion. The independent think time is short enough so that students don't get all the way through the task, but they get ideas started that they can share with their small group or partner. Thus, the small-group discussion becomes more equitable as this process can guard against the fastest thinkers or most eager contributors dominating a group's discussion.

The work continues in small groups, with the teacher circulating and monitoring those groups, asking questions and providing support when needed. Again, it's usually best to stop the groups perhaps a bit short of a fully formed solution. The final phase is a whole-class discussion of different approaches and solution paths. These three components of a lesson, Launch, Explore, Discuss, are integrated within the 5 Practices, and you'll see these phases play out in the description of the remaining practices.

2. MONITORING

 The second practice, monitoring, includes student and teacher routines. Teacher routines include walking around the room, listening to student thinking, looking at student work, asking *purposeful questions*, and *facilitating meaningful mathematics discourse*. Monitoring involves the teacher asking purposeful questions that assess and advance student thinking. Assessing questions center student thinking and encourage reflection, while advancing questions support students in continuing or deepening their thinking in mathematically productive ways.

Student routines include having independent think time to make sense of the problem, choosing strategies, making student thinking visible, sharing ideas and constructing viable arguments with their small groups, and listening to the ideas of peers that expand on their own thinking. This process is best when it's a short enough time that students don't get all the way through the task, but they get some ideas started that they can bring to the small group. The student-to-student small-group discussions are opportunities to build community and student identity. This is an important time in the lesson for the classroom community to cultivate student self- awareness and social awareness as they build relationship skills. As students learn to communicate their ideas and strategies, as well as learn to listen and understand the ideas of others (self-awareness), they are learning to value diverse perspectives (social awareness). Together, these SEL competencies support students in building relationship skills. When students are learning to discuss their ideas, it can be helpful to implement sentence frames to assist with initiating student-to-student discourse during this phase of the lesson.

3. SELECTING

 The process of selecting students to present their work involves the twin considerations of developing a mathematical storyline that advances the understanding of content for all students while attending to the ways in which sharing and discussing student work can position learners as capable of knowing, doing, and communicating about mathematics. The practice of selecting is perhaps the most important of the five in building community. The teacher selects students to present, including their reasoning and sense making, with the key takeaways of the lesson in mind.

> **Tip**
>
> Encourage students to talk to each other about their math strategies, to build a culture that values diverse perspectives and cultivates understanding that there are many ways to solve problems.

There are many ways to build the key takeaways from the lesson; however, if content is the sole driver, we sometimes forget about the importance of ensuring different voices are heard to establish a community of learners. It is important to balance the attention to both the mathematics content and the students. Unpacking the content as part of the process moves us away from the unintentional selection of what might be considered more reliable voices. Disrupting the status quo of hierarchy of intelligence within a classroom, whether it is obvious or not, begins with being intentional in making sure that all voices are valued and that all students are positioned to productively contribute to the discussion. This is accomplished by making sure students have independent think time, then opportunities to share thinking during the monitoring phase, and finally selecting and sequencing student sharing that emphasizes students as problem solvers. The importance of this process cannot be overstated. How others see a student, and how the student sees themself establishes their mathematical identity. This practice supports all students believing in themselves (self-awareness) and disrupts the notion that only some students are born with the "math gene."

> **Tip**
>
> Cultivate a community of learning by encouraging students to share their mathematical ideas during the whole-group discussion. Then students will see themselves and others as valuable contributors, which increases student identity.

4. SEQUENCING AND 5. CONNECTING

The last two practices, sequencing and connecting, are central to how the content will come together during the whole-group discussion. There are various ways to sequence student work, and it is important to make the sequencing and connecting of this work position students as knowledgeable. For example, if the first student work being shared is perceived as the easiest, most common, or most simple strategy, students might begin to think of the first presenter as having less valuable work. In addition to how work is sequenced, the teacher can leverage various strategies being shared by focusing students in the classroom on the connections between the strategies. For example, asking questions such as "How does the diagram in Alex's work connect to the table in Juan's work?" positions both students as having valuable thinking and promotes community within the classroom.

Tips

- Orchestrate student sharing of their work to highlight content goals.

- Broaden student problem solving by having them take on the perspectives of others to build their own understanding.

EXAMPLE OF THE 5 PRACTICES IN PRACTICE: SCOTT'S WORKOUT

Let's bring the 5 Practices routines and our tips to life by considering the Scott's Workout task in Figure 7.2.

FIGURE 7.2. Scott's Workout

Open Up the Math: Launch, Explore, Discuss

Scott has decided to add push-ups to his daily exercise routine. He is keeping track of the number of push-ups he completes each day in the given bar graph, with day 1 showing he completed 3 push-ups.

After four days, Scott is certain he can continue this pattern of increasing the number of push-ups he completes each day.

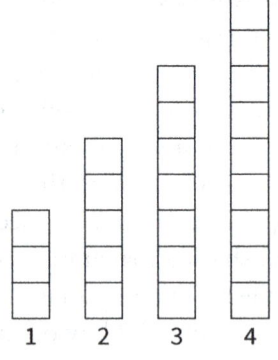

SOURCE: Open Up Resources (2021).

Take a moment to model the number of push-ups Scott will complete on any given day using multiple representations (geometric model, table, graph, equations). Anticipate different ways others may see the pattern grow, and add these to your work. After thinking through the problem, consider the student work in Figure 7.3. What strategies did you anticipate, and which are new? Anticipating strategies becomes easier over time, and when we strengthen our practice of anticipating, we broaden our own understanding while positioning students as mathematicians. We cannot always anticipate every strategy our students will use, and yet when a new strategy is presented, it becomes easier to listen to the ideas of students because we have become more flexible in our own understanding of mathematics. Remember, a lack of anticipation often leads to directing students to a specific strategy rather than to building from students' thought processes. Not listening for understanding of student ideas and, instead, directing them to our way of thinking is a deficit practice. The message students receive is that they are not competent and that their ideas are less valid than our own.

FIGURE 7.3. Scott's Workout Student Work

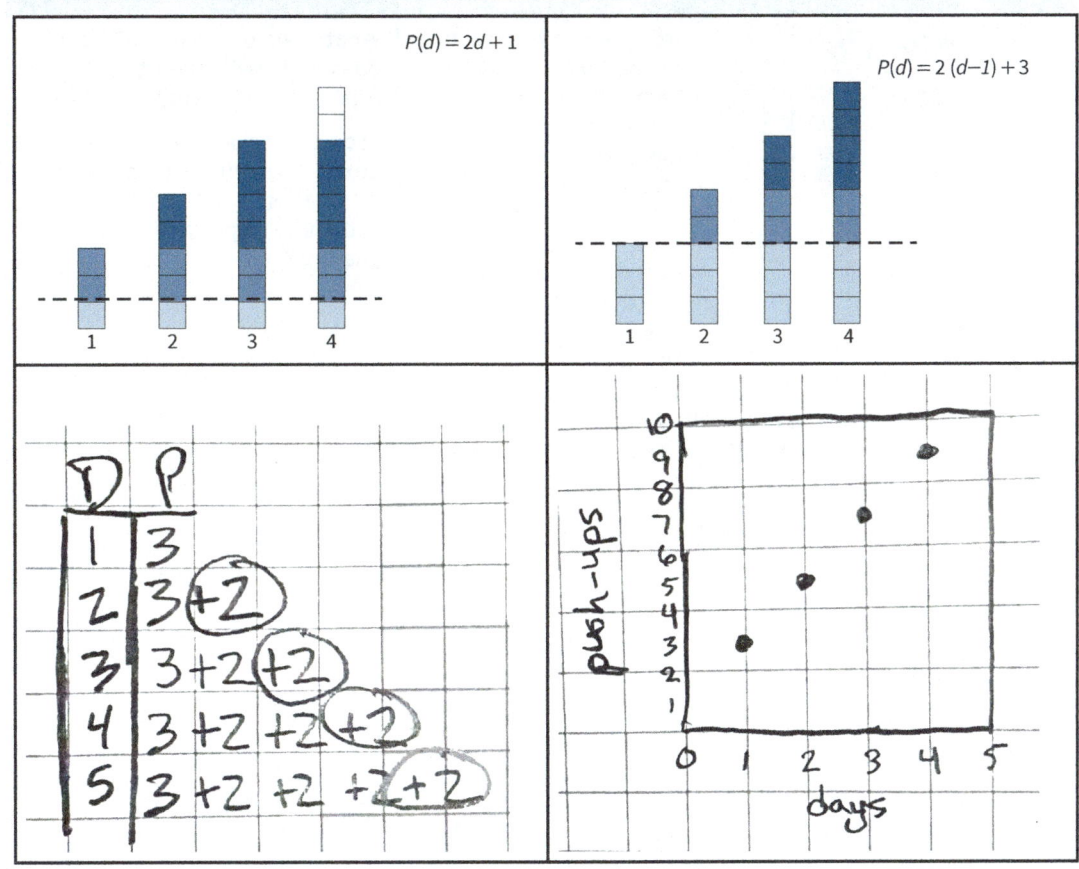

SOURCE: **Open Up Source (2021).**

What might you consider as you engage in monitoring student thinking? Anticipating student strategies is a first step in identifying key questions to ask students as they work. By anticipating, we can create questions to ask during the monitoring phase based on anticipated student thinking. Asking questions to understand students is effective for all students, but it especially supports students who have historically not felt a sense of belonging in their math classrooms. Students who are neurodiverse or learning a new language are supported when they have choice in how they show understanding. Listening to students and using their work shows that we value their thinking and we're more likely to build on the assets they are bringing to the problem rather than redirecting their thinking to the ways in which we as teachers thought about the problem. Table 7.1 shows an example of some anticipated strategies and monitoring questions that teachers might prepare prior to the lesson.

TABLE 7.1 Anticipated Solution Strategies and Monitoring Questions for Scott's Workout

ANTICIPATED SOLUTION STRATEGY		MONITORING QUESTIONS
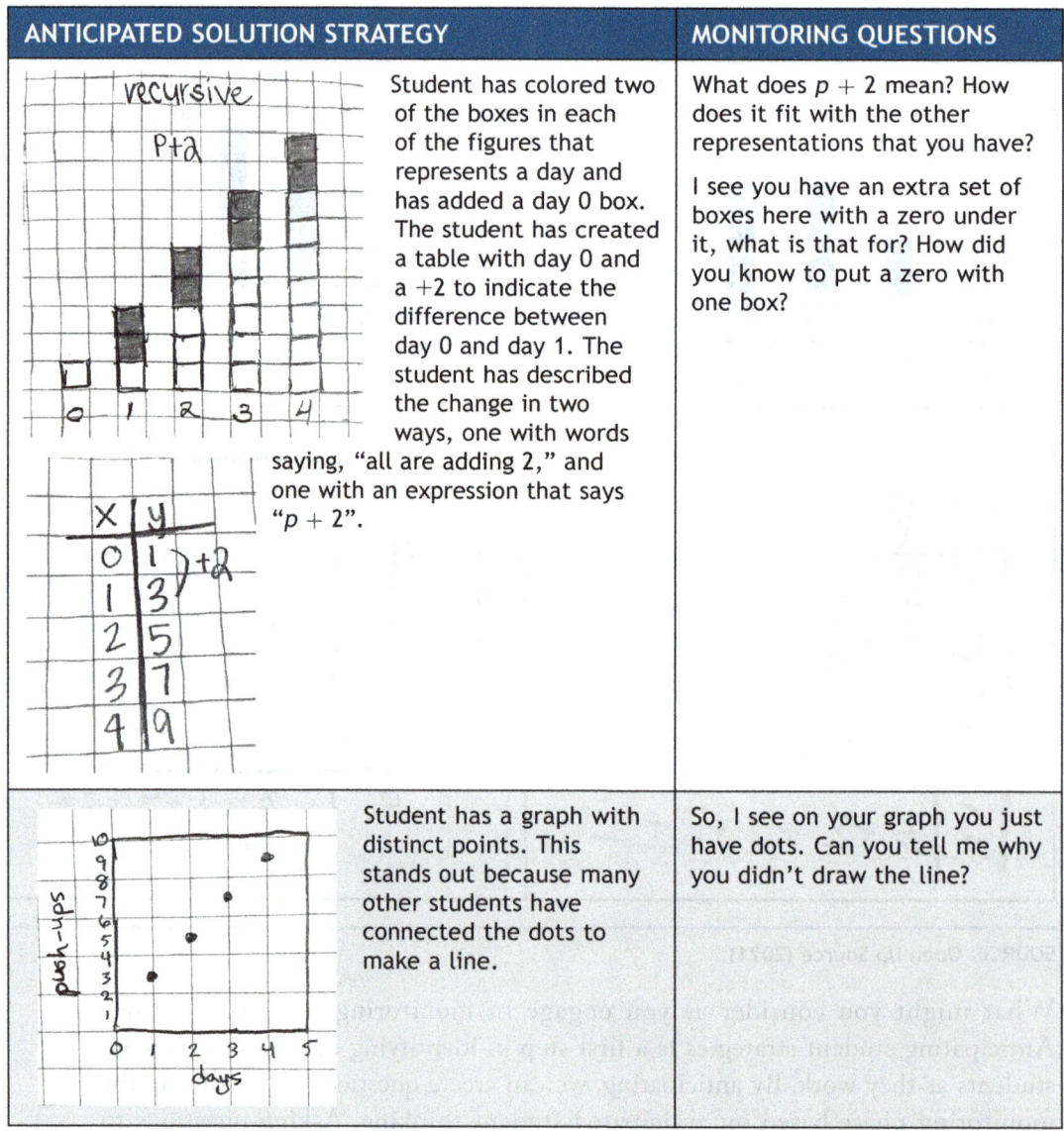	Student has colored two of the boxes in each of the figures that represents a day and has added a day 0 box. The student has created a table with day 0 and a +2 to indicate the difference between day 0 and day 1. The student has described the change in two ways, one with words saying, "all are adding 2," and one with an expression that says "$p + 2$".	What does $p + 2$ mean? How does it fit with the other representations that you have? I see you have an extra set of boxes here with a zero under it, what is that for? How did you know to put a zero with one box?
	Student has a graph with distinct points. This stands out because many other students have connected the dots to make a line.	So, I see on your graph you just have dots. Can you tell me why you didn't draw the line?

TABLE 7.1 (Continued)

ANTICIPATED SOLUTION STRATEGY		MONITORING QUESTIONS
	Student has created an expanded table that shows repeated addition of 2 and writes the relationship between the day and the number of push-ups as $P = 3 + 2(D - 1)$	What have you represented by showing the number of push-ups with a series of +2's? Why did you circle some of them? How did you come up with your equation? Would you explain each piece of it? What does the 3 mean in the context? The 2? The $D - 1$?
Student has colored the boxes in the diagram to highlight the growth. The student has added an extra box with 0 under it, and it is labeled to indicate it is the y-intercept. The student has also created a table and stated slope, along with an expression that is labeled explicit, $2^x + 1$. 		I see you have an extra box here with a zero under it. What does that mean? How did you decide to make it one box? How are you using the diagram to find the table and equation? Explain your equation to me, and tell me about each part. What does the 2^x mean? Why is there a +1? If you substitute in $x = 4$, do you get 9 push-ups?

When the teacher transitions the class to a whole-class discussion of Scott's Workout, the work of selecting, sequencing, and connecting begins. With this task, a teacher might consider starting with a tabular representation that shows the increase of 2 push-ups for each workout day. A teacher might then have groups who iterated and annotated the visual representation to share their solutions, connecting the visual and arithmetic increases of two. Groups with different but equivalent generalizations, such as $2d + 1$ and $2(d - 1) + 3$, could then share their solutions and make connections between the visual diagrams and the table. Finally, a graph could be shown that illustrates the generalization and the increase of two via the slope. During the discussion, the teacher might ask the following questions: "Where do you see the two in the equation? in the table? the picture? the graph? How are the representations the same, and how are they different"?

This quick trip through the use of the 5 Practices with Scott's Workout illustrates how the practices provide opportunities to build intentionally on student thinking. These routines promote an asset-based approach in grounding teacher decision-making in the analysis of student thinking. In addition, the use of the 5 Practices can support the development of SEL competencies from the CASEL Framework, as we share next.

DEVELOPING SEL COMPETENCIES WITH THE 5 PRACTICES

As described in the example of Scott's Workout, during each phase of the 5 Practices, opportunities exist to attend to both the development of the mathematics content and SEL. This example demonstrates how the 5 Practices support the infusion of content and SEL throughout a lesson via teacher moves. These asset-based routines promote a culture of learning and increase student identity and agency in mathematics.

Scott's Workout comes from secondary materials, but the 5 Practices is an asset-based routine that works at all grade levels. Several resources are available to support implementation, including the Five Practices in Practice series (Smith, Bill, & Sherin 2020; Smith & Sherin, 2019; Smith, Steele, & Sherin 2020) and curricular resources such as Open Up Resources (https://openupresources.org/). Figure 7.4 includes sample tasks from other grade bands.

FIGURE 7.4 Sample 5 Practices Tasks

| There are 27 students in our class, and 45 students in the other two second-grade classes. How many second graders are in our school?

(PK-2) | I bought a bag of 48 lollipops to share evenly between 12 people. How many lollipops will each person receive?

(3-5) | Diego has 7 packs of markers. Each pack has x markers in it. After Lin gives him 9 more markers, he has a total of 30 markers. Represent this situation using a tape diagram and one other representation.

(6-8) |
|---|---|---|

Anticipating strategies sets the stage for monitoring students as they work individually and in small groups and for conducting a whole-class discussion that synthesizes ideas. Table 7.2 summarizes how the 5 Practices reflects a set of asset-based routines and actions, resulting in asset-focused student outcomes.

TABLE 7.2   5 Practices as an Asset-Based Routine

PRACTICE	ASSET-BASED ROUTINES	ASSET-BASED ACTIONS	ASSET-BASED STUDENT OUTCOMES
Anticipate	Recognize what is valid or correct in student thinking that may otherwise not be an accurate solution to the task	Builds from what students know as opposed to from what they do not know	Build confidence and persevere in solving problems
	Consider different strategies students may choose	Positions teachers to build on student thinking rather than redirecting students to a specific strategy or the teacher's preferred procedure	Strategies are recognized and valued by peers
Monitor	Orchestrate structure for productive discourse	Implements independent think time for students prior to partner/small-group discussions	Makes sense of the task and identifies entry points/strategies for solving the problem
	Use talk structures to support peer to peer engagement	Creates a community environment where all voices are heard	Cultivates self-awareness and social awareness as they build relationship skills
	Listen to students to elevate their voice	Offers informal assessment of student understanding	Communicates and builds mathematical identities in small-group discussions
	Circulate to observe strategies and ask purposeful questions	Centers teacher-student interactions on student thinking. Supports students in making their thinking visible, constructing viable arguments, and critiquing reasoning of others	Builds mathematical ideas collaboratively and develops self-management skills
Select, Sequence	Use student work to develop a mathematical storyline that attends to content and student mathematical identity	Positions students as mathematically knowledgeable and advances content understanding for all students with the goals of the lesson in mind	Allows for students to know, do, and communicate about mathematics
	Cultivate community of learning	Sequences student sharing in ways that position all students to productively contribute to the conversation and disrupt preconceptions of who is smart in the classroom (Featherstone et al., 2011)	Shows value for the work of all students and how it contributes to the learning community

(Continued)

TABLE 7.2 (*Continued*)

PRACTICE	ASSET-BASED ROUTINES	ASSET-BASED ACTIONS	ASSET-BASED STUDENT OUTCOMES
Connect	Highlight mathematical connections among student strategies Make connections between student work and between concepts	Leverages connections by asking questions such as "How does the diagram in Alex's work connect to the table in Juan's work?"	Promotes community and recognizes students as having valuable thinking as they deepen their understanding of mathematics
	Connect concepts to complete the mathematical storyline	Positions students to identify and record key takeaways	Students co-craft important lesson takeaways and reflect on their learning

Much like aligning a pair of optic lenses increases their power and focus, aligning asset-based language and routines strengthens messages to students about what it means to know and do mathematics. For example, what sense might students make of asset-based language if the daily routines of math class consist of taking notes, watching teachers work examples, and reproducing those examples themselves?

■ ■ ■ Digging Deeper
Designing Routines With Attention to Students Who Are Multilingual, Neurodiverse, or Not Confident in Math

Let's take a look at several groups of students and how the 5 Practices position all students as capable of doing meaningful mathematics, promote an environment where each student feels a sense of belonging, and implement tasks that highlight the relevance of the mathematics being learned in class. Taking particular student populations into account during the design of our classroom routines (both in planning and teaching) can provide broader and more equitable access to diverse students. ■

DESIGNING ROUTINES FOR MULTILINGUAL LEARNERS

As students learn mathematics content, social language, and instructional language, we need to provide opportunities for them to read, write, speak, and listen in their classroom settings. Being intentional in the key use of language supports student development and focuses our work.

In Scott's Workout, students attend to social and instructional language by collaboratively working on the task and then present and explain their work to the class.

Students practice the language of mathematics throughout the lesson. Students explain their understanding of rates of change of linear functions by highlighting the constant rate and making connections between multiple representations. Students listen to classmates' descriptions of different representations, including how the common difference appears in each. Students should learn to explain verbally or in writing (which can be a diagram) how they see a common difference in tables, graphs, and equations. Table 7.3 provides examples of Proficiency Level Descriptors (WIDA, 2020) for various levels of student language development.

TABLE 7.3   Proficiency Level Descriptors With Supports Using Scott's Workout

	LEVEL 1	LEVEL 2-3	LEVEL 4-5
Proficiency Level Descriptors	Identify features revealed in arithmetic sequences written explicitly and recursively with the support of a graphic organizer in a small group.	Describe features revealed in arithmetic sequences written explicitly or recursively with the support of a graphic organizer in a small group.	Argue the benefits of writing an arithmetic sequence explicitly or recursively with the support of a graphic organizer.
Supports	Graphic Organizer (Frayer Model, Concept map, Anchor Chart) Writing Supports (Sentence Frames, Guided Questions) Grouping Structures (Small Group)	Graphic Organizer (Frayer Model, Anchor Chart) Grouping Structures (Small Group)	Graphic Organizer (Anchor Chart, Notes)

As you design lessons, think about where you can incorporate space and explicit attention to the supports that learners who are at different levels might need to make progress. Remember also that many of these supports can benefit all learners, so ensure that all students (not just those identified as language learners) have access to the supports.

DESIGNING ROUTINES FOR STUDENTS WITH DISABILITIES AND/OR STUDENTS WHO ARE NEURODIVERSE

Throughout this book we have been using the UDL Math Framework; however, you may also be familiar with the broader cross-curricular Universal Design for Learning (UDL) Guidelines, from CAST (2018; see Figure 7.5). These guidelines are frequently used by schools and districts as a cornerstone of their planning guidance. We believe the UDL Math Framework provides a more explicit lens for mathematics; however, we also want to include an example of the more broad guidelines from CAST to show the close relationship between the two. The UDL Guidelines offer a set of concrete suggestions that can be applied to ensure that all learners can access and participate in meaningful, challenging learning opportunities.

The guidelines in the UDL Framework incorporate multiple means of engagement, representations, and expression.

FIGURE 7.5. Universal Design for Learning Guidelines

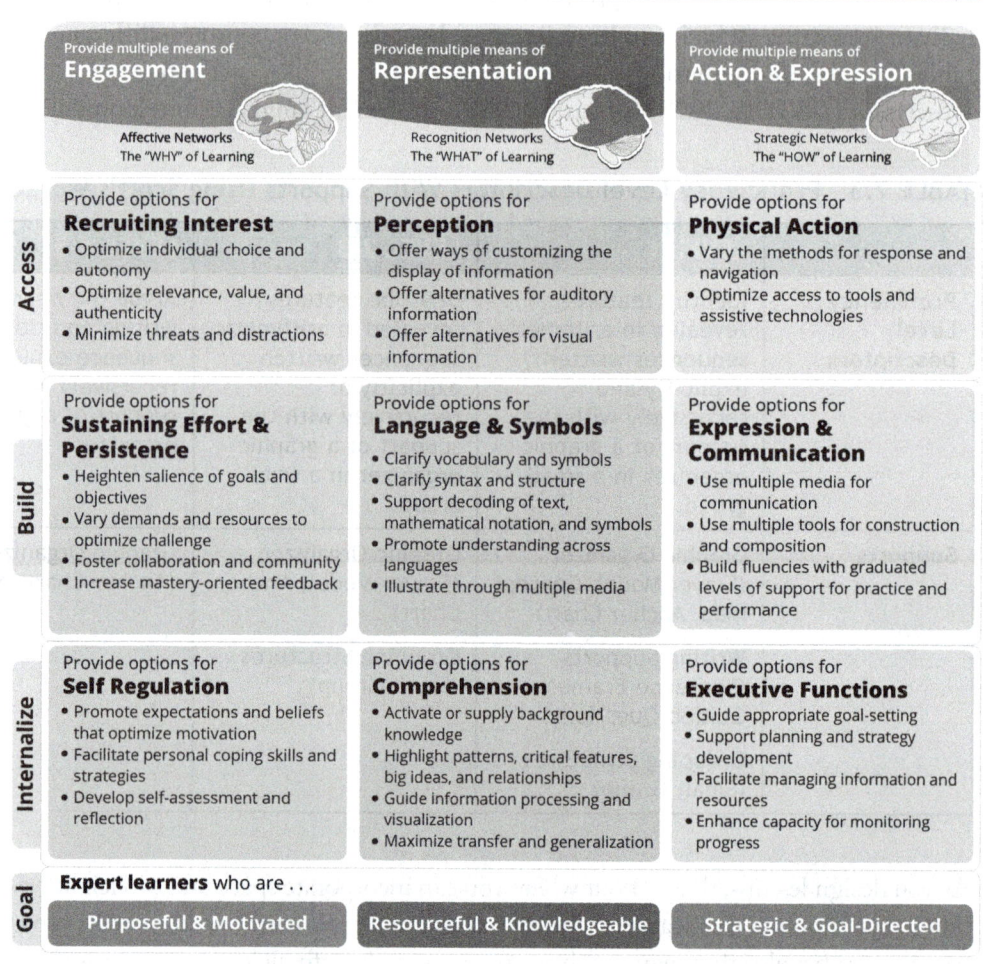

The 5 Practices align with the rows from the framework as both provide access, build knowledge and skills, and incorporate time to internalize learning. The 5 Practices start with mathematical goals and a worthwhile task. The task ensures students have access by including multiple entry points and recruiting interest to the task. Throughout the lesson, students are provided choice and voice in selecting strategies and sharing ideas with peers and the teacher. The sharing of student work during selecting, sequencing, and connecting includes different options for action, expression/communication, including students sharing their work via written text or diagrams, with peers in small groups, with the whole class while standing at the front of the room, or by having students walk around the room during a structured gallery walk.

In the Scott's Workout task, specific aspects are highlighted across the UDL Framework that can support engagement from students with disabilities. Examples are provided in Table 7.4 for each area of UDL: engagement, representation, and action and expression.

TABLE 7.4   UDL Framework and Corresponding Implementation for Scott's Workout

UDL FRAMEWORK	IMPLEMENTATION FOR SCOTT'S WORKOUT
Multiple Means of Engagement	
Recruiting Interest: Optimize individual choice and autonomy	Students use the context and pattern to make sense of the situation and then select representations of their choice to explore what is happening in the problem.
Sustaining Effort and Persistence: Foster collaboration and community	Students begin work independently but then work in small groups to share their work and listen to the ideas of others. Student sharing of unique approaches and viewpoints, and connecting student approaches to the underlying mathematics as a class, promotes collaboration and community.
Multiple Means of Representation	
Provide options for language and symbols: Clarify syntax and structure	Compare the structure of recursive and explicit equations for arithmetic sequences and make direct connections between the structure of the equations different representations (for example, highlight the common difference found in the table and connect this to where it shows up in the explicit and recursive equations).
Provide options for comprehension: Highlight patterns, critical features, big ideas, and relationships	Select student work that exemplifies the relationship between the different representations chosen, highlighting the big ideas of arithmetic sequences.
Multiple Means of Action and Expression/Modes of Communication	
Provide options for executive functions: Facilitate managing information and resources	Provide explicit instructions during the launch for students to create representations of the situation that make sense to them but do not give away the mathematics. Representations may include a diagram, table, graph, explicit formula, and recursive formula.

DESIGNING ROUTINES FOR STUDENTS WHO ARE NOT CONFIDENT IN MATHEMATICS

The 5 Practices promotes the importance of a community of learning, which means that all voices are valued and contribute to the learning of the mathematical goals. The interweaving of the Effective Teaching Practices of the National Council of Teachers of Mathematics (NCTM, 2014) and the 5 Practices (Smith, Steele, & Sherin, 2020) positions students as thinkers and doers of mathematics while maintaining high expectations of deeply learning content. Students are provided independent think time to write and process the mathematics. Then they are encouraged to share their ideas and listen to the ideas of others before the lesson culminates with a whole-group discussion that ends with key takeaways co-constructed between students and the teacher.

In Scott's Workout, students choose *representations* to show how the pattern grows and then explain their solution path to a partner or small group. The small group builds on each other's work and adds to their own work as they learn from one another. Sentence frames are an example of a scaffold that can help support student *discourse* and the transition from small-group to whole-group discussion. The whole-group discussion is orchestrated so that students feel capable of representing an arithmetic sequence and can co-construct the key takeaways of the lesson with peers and the teacher.

PRACTICES THAT BECOME ROUTINES: SUMMARIZING THE ASSET-FOCUSED OPPORTUNITIES

We see several routines embedded in the 5 Practices model. These routines include the practices explicitly named in the model (setting goals and selecting tasks, anticipating, monitoring, selecting, sequencing, and connecting); effective teacher practices such as goals, discourse, and eliciting student ideas; and grouping strategies such as independent work and small- and whole-group discourse. The tips we've shared all focus on how we can hear student thinking, situate it into the flow and goals of a lesson, and encourage other students to build on that thinking in meaningful ways. This approach can help transform the vibe of math class from one in which students receive information that they entered the room not having to one in which they build on and grow each other's ideas into something more powerful. The 5 Practices model is just one example of the ways in which thoughtfully planning using multiple routines together can put student assets at the center of learning during the beginning, middle, and end of a lesson.

Reflect, Apply, Transform

Let's return to where we began the chapter. In the Alignment exercise, you considered a favorite lesson, considered why it was a favorite, and thought about where there were opportunities to leverage student thinking. Now that you've considered the 5 Practices as a set of routines that can frame daily lesson practices, we'd like to invite you to reflect again on the routines that shape your daily practice. The 5 Practices are an example of asset-based routines that, when used effectively, provide students with opportunities to showcase their mathematical thinking, build SEL competencies, and promote positive student identity development. As we have noted using the Practices icon throughout the chapter, these routines can facilitate aspects of NCTM's Effective Teaching Practices.

To close out the chapter, write down as many as you would like of the routines that you regularly use as a part of your practice. For each routine, write down which of the Effective Teaching Practices the routine has the opportunity to build (this might include multiple practices for a given routine). Next, consider the nature of feedback that the routine affords. Do students receive feedback from their peers? From you as the teacher? Does the routine not include explicit or implicit opportunities for feedback? Not every routine may need to provide feedback opportunities, but you may wish to reflect on the balance of feedback opportunities across your regular routines. Finally, note where you feel your enactment of this routine resides on the continuum of deficit-to-asset perspectives, and make some notes about how you might move the enactment of the routine toward a more asset-based perspective.

An example of what this reflection might look like is shown in Table 7.5. Fill out a similar table for yourself.

TABLE 7.5 Reflection Example

MY ROUTINE	RELATED EFFECTIVE TEACHING PRACTICES	OPPORTUNITIES FOR FEEDBACK	ASSET CONTINUUM
Having students write their solutions to example problems on the board	Elicit and use evidence of student thinking	I and other students can share their feedback on the problem and correct mistakes (but this doesn't always happen in every lesson).	X Deficit — Asset The practice problems that I give usually only have one solution and use algebraic representations. I could change the nature of the tasks I use to make this routine more asset based.

(Continued)

MY ROUTINE	RELATED EFFECTIVE TEACHING PRACTICES	OPPORTUNITIES FOR FEEDBACK	ASSET CONTINUUM
Asking students to use tables, graphs, and symbolic representations	Implement tasks that promote reasoning and problem solving Use and connect representations Facilitate meaningful mathematics discourse	I usually give students feedback on their representations as a part of homework. Sometimes we pair up to share their answers with a partner.	Deficit [X] Asset The multiple representations allow students to have multiple entry points. If we did more discussion in class, I might get more information about what students are understanding.
			Deficit [ ] Asset
			Deficit [ ] Asset
			Deficit [ ] Asset

After you've completed your table, reflect on the following questions:

1. In what ways do our routines for planning and teaching mathematics reflect asset-based perspectives?

2. How do our routines resonate or conflict with aspects of asset-based language?

3. How can routines foster SEL competencies and support access for all learners?

4. What aspects of our routines are ripe for change?

Asset-Based Systems

In Part 1, we examined how asset-based language impacts outcomes for students and identified strategies that can transform our classrooms and our schools. In Part 2, we discussed various routines we use in the classroom to deepen our understanding of how these routines fall along the asset-to-deficit continuum. In Part 3, we take a step back and take a closer look at how our system at various levels impact all of us.

Part 3 looks at systems in two different categories:

- Focus on teachers, departments, and school-based instructional coaches: Classroom, School Level, and Mathematics Teacher Educator Program Systems

- Focus on coaches, specialists, and district- and state-level administrators: District, Central Office, and State Office Systems

CHAPTER 8

Structures and Their Impact on Mathematics Learning and Teaching

This chapter begins our exploration into asset-based systems. The classrooms in which we teach are a part of larger systems that include many structures that shape us, our students, our teaching, the classes we teach, and what is valued and honored in our educational community. Examining our language and routines are important pieces of fostering asset-based perspectives, but the systems in which they are embedded can govern and mediate change.

Our goal in this chapter is to examine the structures that make up our systems. Some of these structures are quite overt, whereas others are more hidden. As with language and routines, we'll begin by thinking about how these structures can reflect asset-based or deficit-based perspectives. We'll share some broad categories to help us describe structures and to think about how to foster more asset-based perspectives within those structures.

"Isolation is the death knell of our profession." (Steve Leinwand)

Questions to Consider

1. What aspects of the system work with my classroom efforts to foster asset-based perspectives?

2. What aspects of the system don't support, or actively work against, my classroom efforts to foster asset-based perspectives?

3. Where do I have agency to encourage change to my systemic structures?

Do you ever think about your role in your ecosystem and how your work impacts, or doesn't impact, your school and your community? I (Joleigh) am going to confess that for a long time, I did not recognize how important I was in the system of educating our students. As a mathematics teacher, I focused on what I thought was my role by focusing on my students and their success. I wanted to make a difference so I worked hard, made sure I was prepared, attended professional learning, read research, and learned how to meet the needs of students while maintaining high expectations. I wanted my students to have strong math skills so that they could be successful and achieve their career aspirations. I also wanted them to be critical thinkers and learn to advocate for themselves and others so that they would be productive citizens. I believed my impact centered on the success (and lack of success) of what happened in my classroom. Then one day, a colleague nominated me to be part of a district initiative. And that changed everything.

In just a few months, my mind was open to new ideas and my thought process turned from "What am I doing for students?" to "What are we doing for students, families, and each other?" I started shifting from thinking about my students in my classroom to our students in our school. I came to realize that many of the collaborative practices such as communication, teamwork, and problem solving I was using in my class would not only benefit students in learning mathematics, but they were also important workplace skills. Workplace skills that I wanted to develop for my students also became skills I wanted to enhance for myself and, by extension, for our profession.

Twenty plus years later, I am still on this journey. I spend time, and encourage others to do the same, thinking critically about the structures within our classroom and school. How do we assess our system to identify asset- versus deficit-based structures? What is our process for problem solving, celebrating success, and advocating for change? Are we all positioned from a space of empowerment in our school, or are some voices valued above others? My first shift was in my thinking. Moving from my classroom and my students to our classroom and our students was significant. My biggest shift, however, has been the journey to recognize asset- versus deficit-based structures and to learn how to advocate for change from the position I am in. We are all on a learning journey. I hope this chapter supports you in further recognizing your importance in your role and empowers you to consider actions that would move your ecosystem further along the asset-based continuum.

ALIGNMENT EXERCISE: ASSET-BASED PERSPECTIVES AND SCHOOL STRUCTURES

■■■ Try This

Let's start by looking at some of the more visible structures within our schools. Consider the structures in Table 8.1. For each one, think about what aspects of the structure might align best with asset-based perspectives. Then, think about what aspects of the structure might align with deficit-based perspectives. We've also left a space for you to talk about other reflections or comments about the structure.

TABLE 8.1 Structures and Their Alignment With Deficit- and Asset-Based Perspectives

STRUCTURE	WHAT ASPECTS OF THIS STRUCTURE ALIGN BEST WITH ASSET-BASED PERSPECTIVES?	WHAT ASPECTS OF THIS STRUCTURE ALIGN WITH DEFICIT-BASED PERSPECTIVES?	WHAT OTHER REFLECTIONS OR COMMENTS DO YOU HAVE ABOUT THIS STRUCTURE?
Professional learning communities (PLCs)			
Grades and grading systems			
How students are placed in classes or courses			
How we recognize student performance in mathematics within the community			
How we talk about mathematics in the community (including other adults in our school or district)			

You can think about the structures in a general way, or you can think about the specific ways that the structure exists in your school or district. If you're working with a small group, you might discuss each structure in turn and write down your collective thoughts. ■

We've shared below some thoughts about each structure based on our own experiences teaching and supporting districts. You might think about how these are similar to, and different from, what you wrote.

PROFESSIONAL LEARNING COMMUNITIES

PLCs can be great sources of asset-based perspectives and work. One community that stands out in this way is the PLC at Holt (Michigan) High School, which I (Mike) have written about with a Holt teacher in our book, *A Quiet Revolution* (Steele & Huhn, 2018). The PLC at Holt High School regularly reads the latest research articles and discusses how the articles provide new insights into how students learn. The group also regularly brings student work artifacts to the PLC to talk about what they notice. Common assessments are graded collaboratively, with rubrics carefully designed to identify a wide range of student thinking and reasoning. Not only does this position teachers to identify and honor student strengths, but also it provides teachers opportunities to see the strengths of other students who may not be in their classrooms.

Other PLCs we have witnessed can foster deficit-based perspectives. PLC meetings that focus on what students could not do in lessons are a good example. Although analyzing data can be an important function of a PLC, data dives can sometimes devolve into sorting and categorizing student performance in global and unhelpful ways. In PLCs in which teachers are sharing success stories from their classes, we have sometimes heard other teachers noting that *their* students would not be able to do such things. The reasons behind this are frequently identified as behavioral or related to their prior knowledge or effort in class. Whatever the reason, putting up boundaries about what isn't possible in one's class, no matter the explanatory reason, fosters deficit-based perspectives.

GRADES AND GRADING SYSTEMS

Grades can be powerful motivators for some students. When grades are situated as a part of a system that's designed to provide students with feedback on their performance and pathways to improving and growing that performance over time, they can be very asset-based structures. One way that I used to try to accomplish this in my own classroom was to ensure that grades were provided for a wide variety of performance. Sometimes grades were about correctness, but many other times they were about the mathematical processes and reasoning that was displayed, about student persistence in trying multiple ideas even if they weren't successful, or about collaboration and supporting one another.

But grades and grading systems can also be grounded in deficit-based perspectives. The notion that our grades should naturally have some sort of distribution across the typical letter grades is a great example of a deficit-based perspective. In what ways might this belief guide our hand as we quantify and describe student performance? Would it hold us back from recognizing student strengths? Grading systems frequently serve as gatekeepers to a next course or postsecondary opportunities, sorting those who are permitted to proceed and those who are not. This example also shows the nested nature of our systems—a teacher in a classroom might use grades in a very asset-based way, but the system may use those grades as a sorting mechanism in ways that reflect deficit-based perspectives.

HOW STUDENTS ARE PLACED IN CLASSES OR COURSES

When students choose what they wish to study, and when, they can pursue ideas that are important to them and have a greater opportunity to showcase their strengths. As my (Mike's) youngest child entered high school, they made a choice to take biology and chemistry in the same year as they were particularly interested in science and hoping to pursue a future in astrophysics. Our school system allowed this choice and permitted their to leverage their science assets.

When student agency is not a significant part of the course-selection process, these structures often perpetuate deficit-based perspectives. Often in high school mathematics, courses are rigidly sequential. The opportunity to choose which course you take (for example, when one might take a geometry or statistics course) is not within students' control, and in the rare cases in which such agency is afforded to students, it usually comes through long processes in which students and parents request exceptions to rules. We wonder often what a system would look like that afforded a wider range of choice in mathematics courses to take. My eldest daughter took a course in forensics and my son a course in science ethics. Where are our elective analogs in our math programs?

HOW WE RECOGNIZE STUDENT PERFORMANCE
IN MATHEMATICS WITHIN THE COMMUNITY

You might have approached this structure and thought, recognizing student performance has to be asset-based! We're celebrating student achievements! And certainly recognizing our students for their hard work and performances can be an asset-based practice.

But we invite you to think more deeply about what types of student performances are recognized. Are we focused on students who accelerate into high school courses at the middle school level? Are the students who take Advanced Placement (AP) Calculus and Statistics the ones who are celebrated within the school and in public communications? Are the number of 3s, 4s, and 5s on the AP test featured prominently on the public school dashboard? What students might be systemically excluded from having their accomplishments recognized in mathematics? How might we do this work differently?

HOW WE TALK ABOUT MATHEMATICS IN THE COMMUNITY

Mathematics is often significantly valued in our communities—community members see mathematics as an important subject area and value the production of future science, technology, engineering, and mathematics (STEM) majors in college. Again on the surface, this may seem like it positions mathematics well as an asset in our communities.

But certainly there are deficit-based perspectives when you hear mathematics discussed, as we learned in Part 1. Saying that one isn't good at mathematics in some circles is positioned as a point of pride, a stark contrast to how we might talk about our adult literacy. Student performance is sometimes excused by parents who struggled to be successful in mathematics as if it is a genetic trait that is passed down. As we saw in earlier chapters, these are some more prominent examples of deficit-based perspectives in how we talk about mathematics. But some are less overt and perhaps even more consequential.

One of the most powerful stories in my career came from Craig Huhn, my coauthor in the book mentioned earlier and a teacher at Holt High School. Craig tells the story of walking through the hallways as his prep period started and hearing some of the students he had just completed class with entering another teacher's room. This teacher taught a subject other than math, and they made some offhanded statement that reflected a fixed mindset view of mathematics and a very rote, procedural perspective on what math instruction should be. Craig, as one might expect, had a strong negative reaction immediately. I imagine he fought the impulse to insert himself in the dialogue and try to present a contrasting point of view for the teacher.

But what Craig realized in thinking more about the situation was that neither he nor his colleagues had given other adults in the building an opportunity to learn about their instructional approach to mathematics, to share a dialogue about what they believed about students and their abilities to learn mathematics, and to consider how other teachers might make choices that would support the mathematics teacher team's goals. This story has stuck with me for years. It reminds me that asset-based perspectives exist in complex systems and that we have to provide all aspects of those systems opportunities to learn, change, and grow. That's where we'll take you with the work to come in the rest of this chapter.

HOW DO SCHOOL STRUCTURES REFLECT ASSET-BASED PERSPECTIVES FROM A TEACHER AND A COACH PERSPECTIVE?

Before we dig into the content of the next three chapters, we need to define what we mean by structures. In the context of secondary math education, we use the term *structures* to identify the wide variety of collaborative contexts that make up the schools and districts in which our classrooms are situated. Some of these are visible within our school building, like PLCs and administrative groups. Others operate outside the building, such as district-level administration, state department of education personnel, and groups within the community in which schools and districts are situated. This is a very broad definition, and there are a lot of different facets of systems to cover. In this chapter, we're going to consider systems from the perspectives of teachers and coaches, with Chapter 9 focusing on systems from the perspectives of instructional leaders, district leaders, state leaders, and mathematics teacher educators.

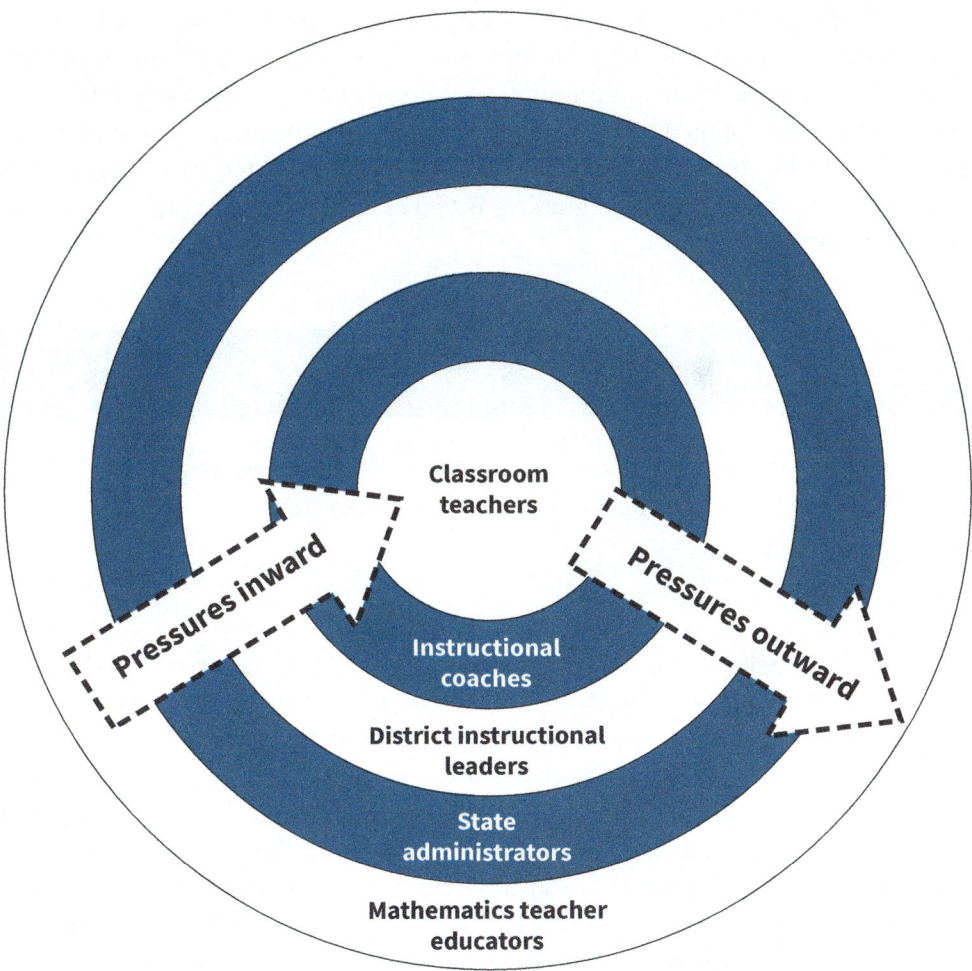

A lot of different structures influence our work as teachers and leaders. Sometimes those structures can resonate with the work we are doing in the classroom to foster asset-based perspectives. Other times, the systems are clearly perpetuating deficit-based messages that may work against our diligent efforts in the classroom. And still other times, whether the structure is perpetuating asset- or deficit-based perspectives is unclear. We contend, however, that there is no "neutral" with respect to our structures. All structures are sending asset- or deficit-based messages to teachers and students.

We're going to consider two broad categories of structures in each of the next two chapters. For each category, we'll consider how these structures influence our classroom practice. We'll also consider how we might be able to influence and shape systems in our respective roles.

POLICIES AND PRACTICES

The policies and practices of a system influence the mathematics to which students have access and how teachers and teaching is supported. Examples of policies and practices include grading policies and requirements, policies relating to how students are assigned with courses, and policies and practices related to assessment. We'll talk about assessment in more detail in Chapter 10—for now, let's consider course assignment and grading as two examples of how policies and practices can shape our systems.

■ ■ ■ Try This

Consider the story that follows. What aspects of this vignette represent asset-based perspectives? What aspects represent deficit-based perspectives? ■

It was coming to the end of the year in Mr. Prairie's 10th-grade geometry class and students were making their schedules for the next year. One student, Bantu, had a difficult start to the year. The memorization that the early part of the geometry course demanded was challenging, particularly since English was not Bantu's first language and the highly technical terms were challenging. But Mr. Prairie and Bantu had worked together to help him focus on connecting ideas such as points, lines, planes, and angles together and to discuss theorems and properties in informal terms. They worked to connect to Bantu's mathematical understandings by using pictures and diagrams to ease the language burden. Although this work was slow, the informal understandings eventually helped Bantu make use of those theorems and properties in writing proofs and justifying thinking. He earned a C in the first quarter, earned a C+ in the second quarter, and was making strong progress in the third quarter of the year with a B.

Bantu was very interested in taking a combined Algebra II and Precalculus course in his junior year to provide him additional options for his senior year, such as

Calculus and Data Science. The school's policy required a teacher recommendation, which Mr. Prairie effusively provided, and a B average in the previous course. Bantu came to Mr. Prairie distraught after learning of this requirement, asking him whether there was enough time left in the year to obtain a B average.

What did you notice in this story? In thinking about Bantu's experience in Mr. Prairie's class, we see a teacher and a student working together to leverage assets. Mr. Prairie worked with Bantu using visuals and diagrams in an effort to provide stronger access to the key ideas in geometry to overcome language challenges. It's likely that the work in Mr. Prairie's class represented asset-based perspectives.

The policy, which seems to have a hard grade-based cut-off for access to an advanced class, represents a deficit-based perspective. The policy was likely well intended when it was written—perhaps to ensure that students who would enter an accelerated course had a track record of mathematics success, perhaps to moderate class sizes, and perhaps to help students make a decision about which class to take. The implementation of this policy, with a firm year-long average, seems to demand some consistent level of performance and doesn't take into account trajectories of student learning like Bantu's. If the policy were to be applied without exceptions, how might this message damage the asset-based work that Bantu and Mr. Prairie worked to foster in the geometry classroom?

■■■ Try This

Consider the story that follows. What aspects of this vignette represent asset-based perspectives? What aspects represent deficit-based perspectives? ■

Ms. Juengel was starting her second year in the classroom and was excited to build on the successes of her first year, along with addressing some of the challenges of that year as well. She had worked hard to get students discussing meaningful mathematics and the adoption of a new curriculum that was student centered, discussion based, and technologically rich was going to help with that. She had developed group norms for productive discussions, took summer professional development that helped her refine her instructional routines, and was looking forward to doing many tasks that featured multiple pathways and opportunities for students to share their thinking.

Where she was challenged, however, was the district's grading policy. She wanted to really emphasize that discussions during class time while working through a challenging problem were valuable and believed that engagement with mathematical processes should be reflected in students' grades. The district's grading framework only listed content standards as standards that could be assessed and required 60% of students' grades to come from summative assessments.

What did you notice about Ms. Juengel's conundrum? I (Mike) recall dealing with a similar situation early in my teaching career when grades were required to be 70% summative assessments. I felt as if I was constantly sending mixed messages by asking students to engage in meaningful mathematics discourse for the bulk of our class time, knowing that their grades would rest heavily on their ability to complete individual paper-and-pencil assessment work. This may be toward the left-center of the deficit-asset continuum; although the policy is not explicitly deficit based, it does inhibit Ms. Juengel's ability to explicitly support a wide range of student assets in her classroom. Even though I suspect that a crafty and ambitious second-year teacher would find innovative ways to work around this policy (probably better than I did at the time!), one wonders what a revision of this policy could be that still emphasizes the importance of summative assessment but better welcomes student assets.

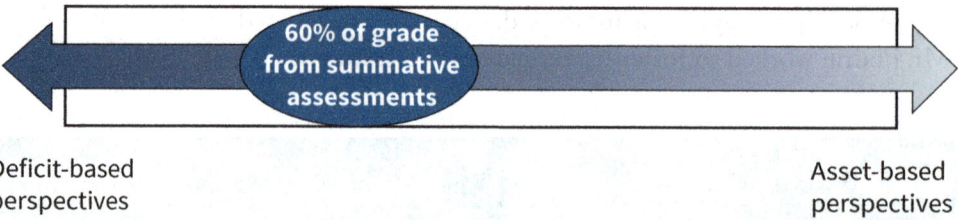

Deficit-based
perspectives

Asset-based
perspectives

As we have seen in these stories, even as classrooms promote asset-based messages, a tracking system that sorts students by perceptions of ability sends a deficit message. The National Council of Teachers of Mathematics (NCTM, 2018, 2020a, 2020b) described the negative impact of tracking on students and noted that tracking itself represents a deficit-based framing of student mathematical learning. Deficit frames play out in tracking not only in how students are assigned to groups or courses but also in how factors like reading fluency or language status create scheduling constraints. We must also examine how systems use and amplify features like large-scale testing scores, public recognition of students for their mathematics performance, extra support for learning, and enrichment opportunities.

Tip

Include a wide range of artifacts in grading for an asset-based classroom, including student discourse, individual work, and group work. This approach acknowledges the range of assets that students bring to the classroom.

An asset-based approach to course selection might instead ask the following: What are the key factors in determining student learning opportunities from year to year? Their demonstrated mathematical capacities and assets, or prior placement in a track or gifted/high-ability label? Are learning opportunities built on research-based learning progressions, or are courses structured based on historical precedent or a curricular resource's content organization? To what extent are systems creating a least-restrictive environment for students? Considering ways to amplify and showcase student performance that reflect student strengths is a critical conversation for creating and strengthening asset-based systems. Diversifying the data sources to measure student progress using student interviews, presentations, and other artifacts that elevate student voice can transform deficit-based systems.

NORMS AND CULTURES

Another facet of systems that is less visible are the norms and cultures of our educational system and the people who make it up. Although these characteristics may be more ethereal, they are no less important to how systems can reflect asset-based or deficit-based perspectives. Understanding the norms and cultures of a system, and how they might change from person to person or group to group within a school and district, can be challenging. And when we uncover an aspect of norms or cultures that we'd like to change, we may not always feel like we have the power to do that as a teacher. The story about Craig Huhn at Holt High School earlier in this chapter is one good example, however, of how we can think about norms and cultures and influencing those as teachers. Let's think through a couple of examples of how norms and cultures might be visible, how they relate to asset-based perspectives on learning and teaching mathematics, and the agency we have as teachers to surface and examine those norms and cultures with an eye towards positive change.

■■■ Try This

Consider the story that follows. What norms or cultures are evident in the story, and how do they foster asset- or deficit-based perspectives? ■

Dr. Walton teaches a Math 3 class for 11th-grade students. Dr. Walton has worked hard to make sure her Math 3 course is welcoming to everyone. Students had discussed it as the hardest course in the school prior to her arrival, and she's tried to take an approach of building on student strengths and providing them multiple entry points into the content of the course, particularly when students perceive that their algebra skills are lacking and getting in the way of thinking through new content. In her district, students have a wide range of choices for a fourth-year mathematics course. They can move to AP Calculus, AP Statistics, Precalculus, Math Modeling, Quantitative Reasoning, or Data Science. Her students met last week with their counselors to put together their schedules.

Freya is a solid student in Dr. Walton's class. She has aspirations to attend college and major either in Spanish or global relations. She has expressed a keen interest in mathematics and wants to take both AP Calculus and Data Science next year. Her counselor noted that her AP Calculus credits probably won't count toward a major in Spanish or global relations and she won't really need those skills for what she wants to do. Her counselor said, "Those courses are really most relevant for people majoring in STEM fields." The counselor strongly encouraged her to think about a lighter load for her senior-year mathematics class.

Tip

Revise course selection policies to include student choice as a significant factor. (Even better, make student choice the *only* factor.)

What did you notice in this vignette? Math 3 (similar to Algebra 2) can often be viewed as a barrier for students and has historically functioned in that way. Dr. Walton's approach to the course explicitly attends to that status and is designed to help students see pathways to success. The guidance counselor in the vignette is approaching advising Freya with the best of intentions—providing clarity about what courses may or may not be most relevant to her intended major. The counselor is also sharing their perception about who belongs (and doesn't belong) in the AP Calculus course.

In this story, we see a clash of norms between Dr. Walton's class and the high school's guidance office. What might Dr. Walton have said to Freya about her intended course-taking options, and how might it be similar to, or different from, the advice she received from her counselor? Different groups within a school and a district may end up sending subtle, and different, messages about who belongs in mathematics (or in specific mathematics learning experiences) and who does not. Sometimes these messages are more overt, like the one in the story. Sometimes they are more subtle, such as a course scheduling matrix that might prevent students from taking an AP Calculus class and a Drama course in the English department. Situations like these can create *de facto* tracking structures even when schools and teachers work to open up students access to a range of mathematics courses. The issue of who has access to what mathematics and when is complicated and nuanced with a wide variety of factors coming into play. But too often, factors like student choice and aspiration are backgrounded, or ignored entirely, in favor of those that feel more quantifiable and easier to justify, such as grades or scheduling constraints.

Strengthening asset-based messages about who belongs in mathematics starts with open conversations about the topic among the various personnel in the school and district. As teachers, we need to listen for stories like Freya's to help us understand what messages are being sent within the district. More broadly, we as teachers can work proactively to help administration and support staff better understand how we'd like to welcome students into mathematics learning experiences, what the barriers might be to access, and how we can work to remove as many of those barriers as possible.

Let's take a look at another scenario that represents the norms and cultures of a school: how success is celebrated.

■■■ Try This

Consider the story that follows. What norms or cultures are evident in the story, and how do they foster asset- or deficit-based perspectives? ■

Mx. Costa is finishing up their first year at South Side High School. South Side takes a very active approach in celebrating student success, with lots of student work hung in the hallways. They also have an annual Community Showcase, where they open the doors of the school to local government leaders and community members and invite them to see what students have produced. The hallways buzz with conversation about student artwork, monitors playing videos of orchestra and choir performances, live science demonstrations, monologues performed by students in the literature and drama courses, and other displays in the hallway.

The math department has traditionally used the Community Showcase to share its success in AP coursework. The department hangs up some posters from projects in AP Statistics and has a few AP Calculus BC students on hand to talk about the colleges they're attending and their interest in STEM majors. Mx. Costa is just finishing teaching the senior-year Quantitative Reasoning class, in which students have been working on projects using functions to compare and make decisions about mortgages, auto loans, and credit cards. They also have routinely discussed *The New York Times*'s What's Going On in This Graph? tasks that ask students to analyze real-world data. Mx. Costa is wondering whether their students would be welcome as a part of the Community Showcase to discuss this work.

This story highlights some very subtle norms and cultures at South Side High School. What gets celebrated publicly about student learning and, specifically, about student learning in mathematics? AP course completion and the spotlighting of students entering exciting and cutting-edge STEM careers are often featured in school and district communications about mathematics achievement. The number of students in those advanced courses and their success rates on AP exams are frequently featured in annual reports and public-facing dashboards. It's worth carefully considering which students these data include and which students these data systematically exclude. Although it may be more challenging to think about how to publicly celebrate mathematics success as compared with, for example, our performing arts colleagues that hold regular public showcases, it is worth thinking about how we celebrate mathematics success and what those celebrations (or lack thereof) communicate about the value of mathematics learning for our students.

Focusing on asset-based perspectives in our classroom is an important step in our students' mathematical learning experiences and the development of positive mathematical identities. But our classrooms do not exist in a vacuum—they are a part of larger, more elaborate systems and both teachers and students need to navigate various parts of these systems every day. When we focus solely on our own classrooms, we run the risk of other parts of the system inadvertently undoing the asset-based work we are doing in our classrooms.

■ ■ ■ Digging Deeper:
The School and District Community

The school and district community may seem like an unusual group on which to focus a digging deeper spotlight. But at the same time, that very fact suggests that this group may be overlooked in the context of mathematics teaching and learning. How can we support the school and district community in asset-focused engagement? This section includes some ideas about community engagement in the context of mathematics, with a focus on asset-based perspectives. ■

WHY FOCUS ON THE COMMUNITY?

Just like the other aspects of systems that we've discussed in this chapter, the community can reinforce asset-based perspectives that are present in your classroom or work against them with well-intentioned, but deficit-based, framing. Consider the cultural norm of proudly saying that one is "bad at math" as one such example. If we're working with students in seeing their mathematical assets and leveraging them every day in class, and these sorts of messages are prevalent in the community, which are students taking away?

Based on our national history in mathematics education, our community members were likely to have experienced mathematics learning differently than the ways in which we are currently trying to support mathematics learning (Larson & Kanold, 2016). Their experiences were more likely to focus on memorization and procedures as compared with meaning-making, and this instruction may have sent messages that speed and recall were key to mathematical success and that you either knew math or you didn't. The approach to mathematics learning that we and others like NCTM (2014, 2018) are promoting, with a strong research base behind it, is very different. We need to work with our community members to understand that.

An implication of this situation is that community members may not have ever been asked about how they use their own mathematical assets. They might not even be keenly aware of those assets! We know that mathematics is embedded in a wide range of tasks across a range of professions, and asking for community members to think about the mathematical assets they use every day can create powerful allies in mathematics teaching and learning. The strategies we share below provide opportunities to challenge deficit-based norms, cultures, and beliefs and support community members in identifying the mathematical assets that they use.

COMMUNITY MATH NIGHTS

Community math nights are more common in the elementary grades, but there is absolutely no reason not to hold community math nights in middle and high school contexts. In these events, engage community members in rich mathematical tasks in ways that we would engage our students, promoting discussion, problem solving, and building on their mathematical assets. Opening with a math activity sets important context as it helps community members understand what contemporary mathematics instruction looks and sounds like, how it may be different from their own learning experiences, and provides an entry point for deeper conversations about mathematics teaching and learning. A discussion-based task like Scott's Workout (featured in Chapter 7) or brief discourse-based tasks like Which One Doesn't Belong are great choices as centerpieces for a community math night.

> **Tip**
>
> Let students serve as facilitators for math tasks during community math nights.

From this jumping-off point, community math nights can showcase students work, create opportunities for community members to engage with students around mathematical projects that they are working on, or dig more deeply into conversations about curriculum and instruction. They can also be an opportunity to share with the community the spectrum of mathematics courses students can take in middle and high schools and discuss the goals of those courses.

ASSET-BASED MATHEMATICS COMMUNITY DISCUSSIONS

Another way to engage the community is to develop community asset-based mathematics discussions. These events can be synchronous or asynchronous, but the audience should be secondary mathematics classes. At these events, community members across a range of professions are invited to share how they use mathematical assets in their professional work. In addition, community members can be invited to share what aspects of mathematical and statistical thinking they wish they had the opportunity to learn in secondary grades but have instead learned in their careers.

Think of this event as the next iteration of a career night, in which professionals talk about their jobs. Those events typically were designed to stimulate students' interests in various careers. An asset-based mathematics community discussion has a different aim—to show how aspects of mathematics thinking are used across a wide range of careers. This distinction may seem minor, but it is important.

In developing such a conversation, recruit from a wide range of career opportunities in your community. We encourage you to meet with the career group beforehand to share a bit about what we mean by an asset-based perspective on mathematics and to help them calibrate their remarks to reflect an asset-based framing. It can also be a great event for teacher professional development. For example, I recruited the data analytics group from Major League Baseball's Milwaukee Brewers baseball club to speak to teachers about how they use mathematical and statistical thinking in their work. These discussions gave teachers valuable insights into how to foster real-world problem solving with their students.

Reflect, Apply, Transform

Return to the structures in the Alignment exercise, and select one to focus on (or propose a new one of your own). Choose a structure about which you feel the most passionate. List some aspects of how the structure currently works. For example, if you choose how students are placed in courses, list the criteria and process currently used. Place each of those aspects on the deficit-to-asset continuum.

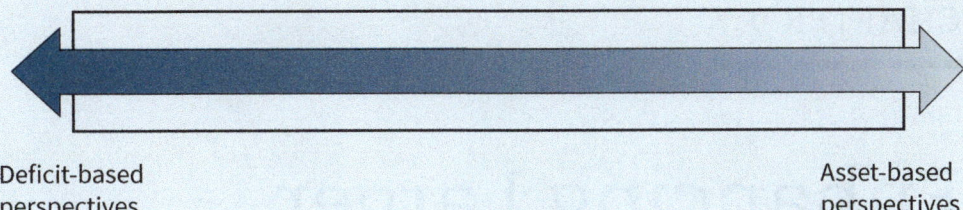

Deficit-based
perspectives

Asset-based
perspectives

Next, identify the constituencies that have voice in this structure: students, mathematics teachers, other adults in the building or district, families and community members. Describe specific actions that should be taken (by you, the reader, as an individual or by a collective group) that can move the structure more toward the asset side of the continuum.

This exercise is designed to help you identify the pieces of the structure that are valuable and worth retaining as they reflect asset-based perspectives and the aspects of the structure that are ripe for change. This exercise also helps you identify the allies you might recruit to support change for this structure within the system. As you complete the Reflect, Apply, and Transform exercise, revisit the following questions:

1. What aspects of the system work with my classroom efforts to foster asset-based perspectives?

2. What aspects of the system don't support, or actively work against, my classroom efforts to foster asset-based perspectives?

3. Where do I have agency to encourage change to my systemic structures?

CHAPTER 9

Enhancing Larger Systemic Structures Through Asset-Based Perspectives

This chapter continues our considerations of asset-based education systems. In our next discussion, we will zoom out to view larger systemic structures, analyze where these structures fall along the asset-based learning environment continuum, and then identify how we can foster more asset-based perspectives within those structures. Larger systemic structures include school-wide, central office (district or county offices), and state agencies.

> *"How do our current actions demonstrate our commitment and sense of urgency to improve outcomes and raise expectations?" (Glenna Gallo, U.S. Department of Education Assistant Secretary for Special Education and Rehabilitation Services)*

Questions to Consider

- What aspects of the system cultivate asset-based perspectives?
- What aspects of the system don't support or actively work against efforts to foster asset-based perspectives?
- Where do I have agency to encourage change to my systemic structures?

SOURCE: iStock.com/bernie_photo

In Chapter 8, I (Joleigh) shared my journey and how I wasn't aware of my importance or impact as a classroom teacher for several years. I collaborated with peers and shared ideas. However, I didn't truly see that the work happening within my four walls was an equity issue if our system wasn't working together to ensure all of our students received the best education possible. It wasn't until I took on the role of district specialist that I recognized specific structures within our system could make a big difference.

For example, schools with asset-based professional learning communities (PLCs) resulted in a team working to increase student outcomes instead of one teacher in isolation. Throughout the years, research and working with teachers gave me new perspectives on several structures in our system. What I have learned from myself is first to come to understand why a structure is in place before looking at it with a critical eye. Then, unpack the structure to identify who it is serving and who it is not. How does the current structure best align with asset-based perspectives, and in what ways are there factors of the structure that result in deficit perspectives? Looking at structures and how they impact our system is still a work in progress for me. Currently, I am working with several colleagues to look at Free Appropriate Public Education (FAPE) under the Individuals with Disabilities Education Act (IDEA) of 1990 (reauthorized 2004) to unpack aspects of a student's Individual Education Program (IEP). What is intended by the IEP? What structures support the intention? What structures get in the way? What elements of many implemented IEPs are asset based, and what elements are deficit based? How do we do things differently so students who are identified with disabilities under IDEA achieve at the same level as their peers without disabilities?

What structures are you currently analyzing that impact your system? What structures would you like to examine more closely? Let's look together at various structures in the Alignment exercise.

ALIGNMENT EXERCISE: ASSET-BASED PERSPECTIVES AND SCHOOL STRUCTURES

Education is complex, and most policies, procedures, guidelines, and practices are created with good intentions. However, due to the complex nature of our system, some procedures and practices result in structures that align with asset-based perspectives in one aspect and deficit-based perspectives in another. Let's look at some of the structures of our system and unpack how we attend to these in our work. The goal is to enhance our assets and, where needed, to change aspects that undermine our system.

■■■ Try This

Consider the structures listed in Table 9.1. For each structure presented in the first column, what structure elements might align best with asset-based perspectives? What elements align with deficit-based perspectives? We've also left a space for you to discuss other reflections or comments about the structure plus a couple of additional rows to add other systemic structures you would like to analyze.

You can think about the structures in a general way, or you can think about the specific ways that the structure exists in your school or district. If you're working on this exercise in a small group, you might discuss each structure in turn and write down your collective thoughts.

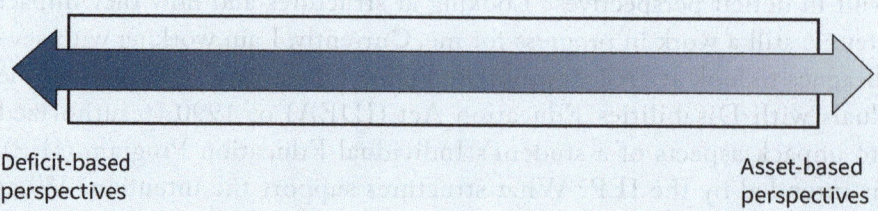

Deficit-based
perspectives

Asset-based
perspectives

TABLE 9.1   Structures and How They Align Along the Deficit-to-Asset Continuum

STRUCTURE	WHAT ASPECTS ALIGN BEST WITH ASSET-BASED PERSPECTIVES?	WHAT ASPECTS ALIGN WITH DEFICIT-BASED PERSPECTIVES?	ADDITIONAL REFLECTIONS OR COMMENTS ABOUT THIS STRUCTURE
Professional Learning (PL)			
Tracking: Students, Teachers, Schools			
Teacher Feedback and/or Evaluations			
Celebrations and Success stories			
Pacing Guides			
Additional topics to consider:			
TOPIC	WHAT ASPECTS ALIGN BEST WITH ASSET-BASED PERSPECTIVES?	WHAT ASPECTS ALIGN WITH DEFICIT-BASED PERSPECTIVES?	ADDITIONAL REFLECTIONS OR COMMENTS ABOUT THIS TOPIC
Individual Education Programs (IEPs)			
Mathematics Education (Beliefs about mathematics)			
Classroom Walkthroughs			
Decisions about mathematics education			

We've shared some thoughts in the following discussion about each structure based on our own experiences plus the experiences of others. How are these reflections similar to and different from what you wrote?

PROFESSIONAL LEARNING

SOURCE: iStock.com/kali9

What are your experiences with PL? Are you aware of opportunities for PL that meet your needs and increase your capacity as a professional? Are there ways that your input is encouraged and considered before, during, and after the PL process? Can you state the learning intentions, success criteria, and the value of the PL? Our time is precious as mathematics educators, and PL can fall in many locations across our deficit-to-asset continuum. Table 9.2 demonstrates a few examples of asset-based and deficit-based traits of PL. There are extra rows for you to consider and add your own examples. In the table,

- participants are those receiving the training,

- organizers are the ones handling logistics,

- facilitators are those who orchestrate the session or sessions, and

- all members are involved in the PL.

TABLE 9.2   Examples of Asset-Based and Deficit-Based Traits of Professional Learning

ASSET-BASED TRAITS OF PL	DEFICIT-BASED TRAITS OF PL
Incorporates ongoing contributions from all members before, during, and after the PL	Assigned PL in which one primary member or group makes decisions with little-to-no input from the other member or group groups. Facilitator is the holder of the knowledge

TABLE 9.2 (*Continued*)

ASSET-BASED TRAITS OF PL	DEFICIT-BASED TRAITS OF PL
Includes opportunities to reflect and listen to the ideas of others	Focus is on information dissemination, with little interaction from participants
Clear goals are established and shared with all participants	Unclear goals or lack of coherence that leave participants feeling confused or unsure of next steps
Outcomes are aligned with goals. All members involved with PL are asked to self-assess growth and identify next steps for learning and implementation	Participants attend and hopefully learn something that may or may not impact their learning as a professional
Aligned with other asset-based structures/goals in the system	PL does not consider (and may even conflict) with other asset-based structures/goals

■■■ Try This

Which asset-based traits in Table 9.2 resonate the most with your desires for effective professional learning and why? What other traits would you add to the asset-based column and to the deficit-based column? How do these traits align with Social Emotional Learning (SEL) for us as adults and professionals? Use Table 9.3 to share your thoughts and comments.

TABLE 9.3 Your Thoughts on SEL and PL

SEL	HOW DOES SEL SHOW UP FOR US AS ADULTS WHEN IT COMES TO PL?
Self-Awareness	
Self-Management	

(*Continued*)

TABLE 9.3 *(Continued)*

SEL	HOW DOES SEL SHOW UP FOR US AS ADULTS WHEN IT COMES TO PL?
Responsible Decision-Making	
Social Awareness	
Relationship Skills	

- As a participant, how can you advocate to enhance PL practices to move further along the asset-based continuum?

- If your role is to support professional learning, identify one current practice of PL in your system that is already asset based and think about possible next steps to move this practice further along the asset-based continuum.

- Identify any areas that are deficit based and create a plan of action to change the current practice. ∎

TRACKING

Chapter 8 addressed considerations for how students are placed in classes or courses—aspects of systems that classroom teachers may have more direct input on and control over. In this chapter, let's consider teacher assignments within and across schools. The National Council of Teachers of Mathematics' (NCTM, 2018) *Catalyzing Change* publication did an excellent job of naming teacher tracking as an essential practice to examine and consider. How teachers are assigned to courses is important and consequential. There are many aspects to consider, with a few of them being:

- Who (and what) determines course assignments?

- How do the assignments allow for collaboration within the department (or across departments)?

- In what ways are students considered when creating the schedule?

- At the district level, what data do we use to ensure teacher tracking doesn't happen across the district (i.e., are most beginning teachers in the lowest performing schools)?

- At the district/state level, how often are Equity Data Dives done to identify discrepancies (course placement, school placement, etc.), and what is in place to ensure our system enhances our profession and outcomes for all students?

When I (Joleigh) started teaching, I was excited to be in a Title I Junior High. I got my schedule (predetermined) and worked hard at all the things I thought I knew. There was a LOT I did not know, but there was also a lot *I didn't know* that I didn't know. For example, I was the only teacher teaching the classes I was assigned. And it wasn't until halfway through the year that I realized my entire schedule was made up of classes in which every student in every class failed math at least once and/or had been placed in the "lowest" level mathematics course for who knows how long. Who helped me learn this? My students.

Tips:

- Increase buy-in by creating a needs assessment (or short survey) to be completed by all participants to identify the learning needs of those involved in the PL.

- Include clear learning intentions and success criteria that align with goals and outcomes at the beginning and then provide time for engagement, reflection, and action throughout the PL.

- Position participants as contributors whose voice is valued, similar to what we want teachers to do for students. This approach includes identifying individual participant strengths and using these to build capacity and leverage your system goals.

SOURCE: iStock.com/BlessedSelections

Another example I didn't know was that I wasn't a great teacher. I did what I thought I was supposed to be doing, which included calling parents, both to praise what was going well and to express concerns, such as students not completing assignments. To this day, I remember calling a parent to ask how I could support their child in getting their homework turned in, and the parent said, "Ms. Honey, you called me about this two weeks ago. I appreciate your call, but perhaps you could place your efforts elsewhere." It took me a while (several years, in fact) before I understood. My students first needed to know that I cared about them. My efforts on "doing it right" missed the importance of SEL skills we have discussed throughout this book. I needed to focus on building relationships, cultivating a community of learning, and developing my pedagogical content knowledge.

> **Tip**
>
> Create an agreed-to structure that includes choice and voice in determining course load. For example, is there agreement about the number of courses or assignments of each teacher? Do all teachers have an opportunity to teach at least one of their courses of choice? How does the schedule impact this discussion?

I share this experience because this discussion isn't about the schedule being inappropriate for a first-year teacher and thus unfair to me (or thousands of other first-year teachers) but that the schedule was not appropriate for the students. Students deserve a great math teacher with expertise and the knowledge that comes with experience and reflection. New teachers deserve opportunities to collaborate and share ideas with colleagues. I am grateful I didn't leave the profession after that first year because my career has been very rewarding. But that first year was difficult.

TEACHER FEEDBACK AND EVALUATION

What teachers are evaluated on, how that evaluation process operates, and how teachers receive feedback in general can reflect asset-based or deficit-based perspectives. Rubrics like the Instructional Quality Assessment, used in productive ways (see Candela & Boston, 2022), can help teachers identify and reinforce productive practices meaningfully linked to student learning.

Some elements of teacher evaluation systems, however, can be strongly deficit based. For example, one criterion I (Mike) was evaluated on was whether I wrote a lesson objective on the board when my principal walked through the room. While *establishing mathematics goals for learning* is one of the NCTM Effective Teaching Practices, just looking for written evidence of a goal doesn't speak to the quality of that goal, how that goal is reflected in instructional practices, or the extent to which the goal is visible and tangible for students. Being evaluated in this way did not provide me with any constructive feedback on my teaching, and it ignored the fact that I might communicate my learning goals with students in a wide variety of ways.

Are there currently aspects of the teacher feedback loop that focus on teacher reflection based on the identified goals of the teacher? How are we, as educators, encouraged to reflect on our own self-awareness, self-management, or social awareness? How do we currently reflect on our decision-making process and how we build relationship skills?

We recommend reflecting on the current practices for feedback loops or evaluations. Consider aspects that could improve and empower those receiving feedback by adding more SEL.

CELEBRATIONS

Similar to Chapter 8 (where we ponder what types of student performances are recognized), we invite you to think more deeply about the types of recognitions across the district and/or state. Is mathematics celebrated? Are team efforts recognized? Do the recognitions celebrate mathematics teaching and learning in ways that are inclusive or exclusive? Do the recognitions include opportunities for students to celebrate and experience the joy, beauty, and wonder of mathematics?

Tip

When creating evaluation rubrics for classroom observations, include questions that promote self-awareness, self-management, and social awareness. Questions can be direct: "How did your self-awareness, or understanding one's own emotions, thoughts, and values and how they influence behavior (Collaborative for Academic, Social, and Emotional Learning [CASEL], 2024), show up in the lesson?" Questions can be open to any aspect of SEL: "Share what you are working on or paying attention to as it relates to SEL. What are your current strengths and what are you working on?"

SOURCE: iStock.com/courtneyk

Make a list of all recognitions/celebrations you are currently aware of (start with math at the school, district, mathematics organizations, and state level and then include other recognitions/celebrations about education). How are they structured to highlight the different strengths and contributions of various members? ■

Tip

Schedule time for recognitions, large and small. This recognition may include nominating colleagues for positions on state and national organizations, nominating teachers for the Presidential Award for Excellence in Science and Mathematics Teaching (PAEMST), selecting colleagues for a monthly teacher spotlight, or a thank-you note for the time they spend tutoring students during lunch.

Tip

Use social media to highlight assets and enlarge mathematics education and mathematics educators.

Just like with students, we must recognize and celebrate the work of educators. And we must identify and celebrate diverse contributions and ensure these are communicated across various platforms, as appropriate (school and community newsletters, faculty and department meetings, social media, email, etc.).

The deficit side of recognition comes in various forms. Many things in our systems can be improved, and we must always be committed to learning and working to do better. Unfortunately, what often gets attention is all the things wrong with our system, our classrooms, our teachers, and our organizations. We encourage you to consider the assets of the situation (or the human) before complaining about what is inadequate, especially on social media. Considering the assets means recognizing what efforts have been made and building from there. It doesn't mean only speaking of the good. Therefore, when we start with what is good, it shows others that we have taken time to recognize their efforts and that we come from a place of good intentions. When we do this, others are more open to us bringing to light what needs to be seen. Crucial conversations are needed for all of us to do better, but when we start with complaints, it may be that we are showing what we don't know more than showing that we care.

PACING GUIDES

Pacing guides are a great example of a practice created to support schools and teachers to ensure all students receive instruction on core standards regardless of which school they attend or which class they are enrolled in. Unfortunately, pacing guides (or curriculum maps) often specify the time spent on a topic at a microlevel. The intention may be to support students and teachers; however, the results of typical pacing guides include heightened pressure for both teachers and students and a more teacher-centered learning environment (David, 2007). When teachers are pressured to "get through lessons," there is more likelihood that our classroom becomes more teacher telling and less student thinking. Sometimes, teachers also skip a lesson to keep up.

		Jan. 2-Jan. 5	Jan. 8-Jan. 12	Jan. 16-Jan. 19
Notes				
Math				
Reading				
Phonics				
Language				
Writing				
Science				
Social Studies				
Chapter Book Read Aloud				

Asset-based pacing guides are co-crafted by those using them, focus on major concepts students learn in a large-ish chunk of time, and are revised periodically by the team of teachers, coaches, and instructional leaders. The major concepts or standards drive the work, not an end-of-year assessment. The team also works together to discuss the takeaways of the conceptual and procedural understandings of what we want students to know and be able to do.

The following story has been created from interviews and experiences with teachers and district leaders from several districts. As you read, consider how the pacing guide in this situation contributes to both asset-based and deficit-based perspectives.

Shannon, a district leader at Smithfield Unified, is part of an eighth-grade team PLC. The weekly meeting starts by discussing where they are in the pacing guide and the four questions from the DuFour model (Dufour, 2021): What do we want our students to learn? How will we know if each student has learned it? How will we respond when they don't learn? How will we respond when they already know it? The teachers have come prepared with their pacing guide and list of standards covered in the unit. The discussion includes what lessons and concepts will be covered. When the conversation shifts to questions about student learning, the focus is on specific skills students should be able to do. Once the skills are listed, the group discusses students who are struggling based on the most recent common assessment data. Participants list who may need interventions to get them caught up. Some teachers note that there were some parts of the assessment in which almost all students did poorly, so the group decides that it will focus on the data from that assessment when it meets next week.

What aspects of the pacing guide are supporting the work of this team? What aspects of the pacing guide may be hindering the work of this team? What ideas could this group consider to adjust its pacing guide and PLC time that would move its work further along the asset-based continuum?

HOW DO DISTRICT AND STATE STRUCTURES REFLECT ASSET-BASED PERSPECTIVES?

In Chapter 8, we defined structures as a way to identify the wide variety of collaborative contexts that make up the schools and districts in which our classrooms are situated. Some of these are visible within our school building, like PLCs and administrative groups. Others operate outside the building, such as district-level administration, state department of education personnel, and groups within the community where schools and districts are situated. This definition is broad, and there are a lot of different facets of systems to cover. In this chapter, we will look at systems from the perspectives of instructional leaders, district leaders, state leaders, and mathematics teacher educators.

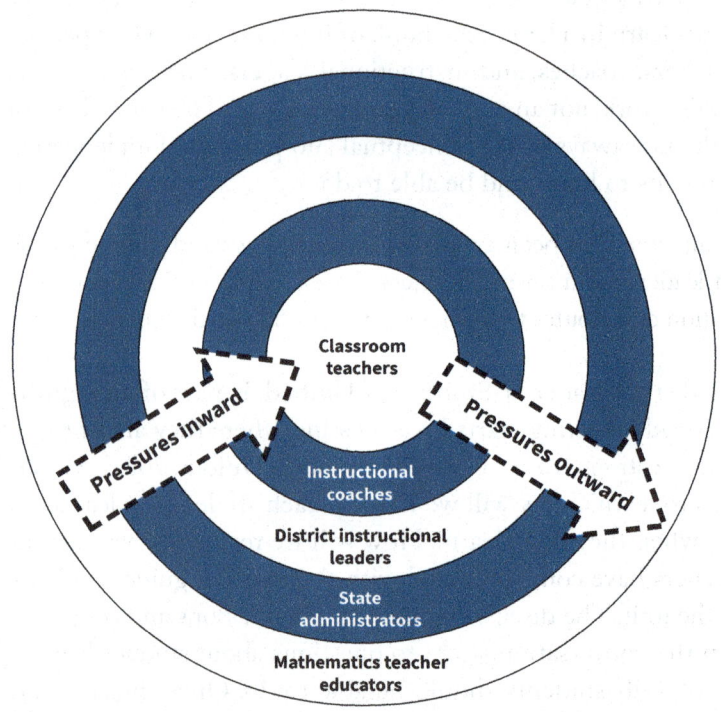

TRANSFORM YOUR MATH CLASS USING ASSET-BASED TEACHING FOR GRADES 6-12

A lot of different structures influence our work as instructional leaders and as district and state employees. Many structures within our system, especially when created by participants at all levels, are asset based. However, although all structures are intended to be asset based, education is complex work, and some things initiated to support teachers and students can unintentionally backfire and have deficit results. Let's continue looking at some of our other systemic structures.

INDIVIDUALS WITH DISABILITIES EDUCATION ACT (IDEA) AND INDIVIDUAL EDUCATION PROGRAMS (IEPs)

IDEA is an example of a federal law that has generated many positive outcomes and has established rights for people who have been identified as having a disability. IDEA is a law that provides FAPE and ensures special education and related services to eligible children with disabilities nationwide. Initially, this law was signed under Gerald Ford in 1975 as the Education for all Handicapped Children (EHA). In 1990, it was reauthorized under George Bush, and part of the reauthorization included the name change to become the Individuals with Disabilities Education Act (IDEA). IDEA was most recently reauthorized in 2004.

Before IDEA (and only ~50 years ago), many individuals who were identified as having a mental illness or intellectual disability lived in state institutions with restrictive settings. At this time, most families were not provided access to resources or opportunities to plan or make placement decisions regarding their children. Today, most children attend their neighborhood schools rather than separate schools or institutions. More than 66% of children with disabilities are in general education classrooms 80% or more of their school day (IDEA, 2004). Other accomplishments directly attributable to the IDEA include improvements in the rate of high school graduation, postsecondary school enrollment, and postschool employment for youth with disabilities.

This law has many policies, procedures, and practices in place at the federal, state, district, and school level. All policies, procedures, and practices must align and comply with IDEA. Perhaps the most known aspect of IDEA is the IEP that is required and is the right of every public school child who receives special education and

SOURCE: USDOE.

https://sites.ed.gov/idea/

related services under this law. According to IDEA, IEPs must include a statement of the child's Present Levels of Academic Achievement and Functional Performance (PLAAFP), measurable annual goals and how goals will be measured, services, and transition plans (starting at age 14). Additional components may be included at the state or district level to support compliance of IDEA or other state policies. What is your current level of knowledge related to IEPs? How can we better support implementation of IDEA by understanding each component of an IEP?

■ ■ ■ Try This

Answer the following questions about IEPs in your sphere of influence and then use the continuum to mark aspects of your current processes and structure as they fall along the deficit-to-asset continuum.

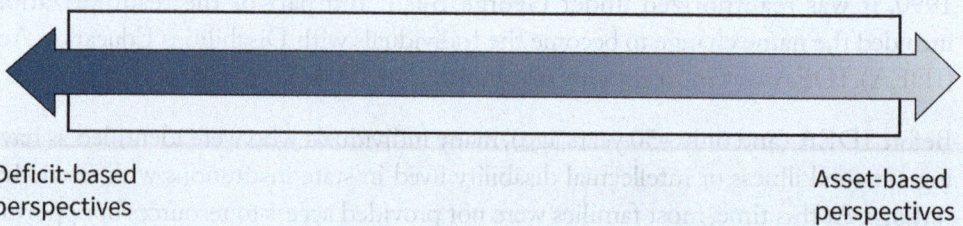

Deficit-based
perspectives

Asset-based
perspectives

Here are some questions to think about to identify aspects of the current structure of how IEPs are created and used:

- Who attends the IEP meetings?

- Who contributes to the writing and revision of the document, including the PLAAFP, goals, transition plans, and services?

- How are student goals determined, and how often are they reviewed or revised?

What aspects of your current processes lend themselves to asset-based perspectives? Deficit-based perspectives? What are your next steps for collaborating with peers to unpack the processes to make the IEP more asset based while still meeting compliance? How do you modify current situations so that compliance and best practices are in alignment with one another? ■

If you would like more information about IEPs, the Digging Deeper section of this chapter takes a closer look at the different components and provides guidance to advocate for better. What actions can be taken to shift our work to be more asset based? It is impressive to see how far we have come during the last 50 years in recognizing students with disabilities, as capable of achieving, but imagine where we *could* be!

HIGH SCHOOL PATHWAYS

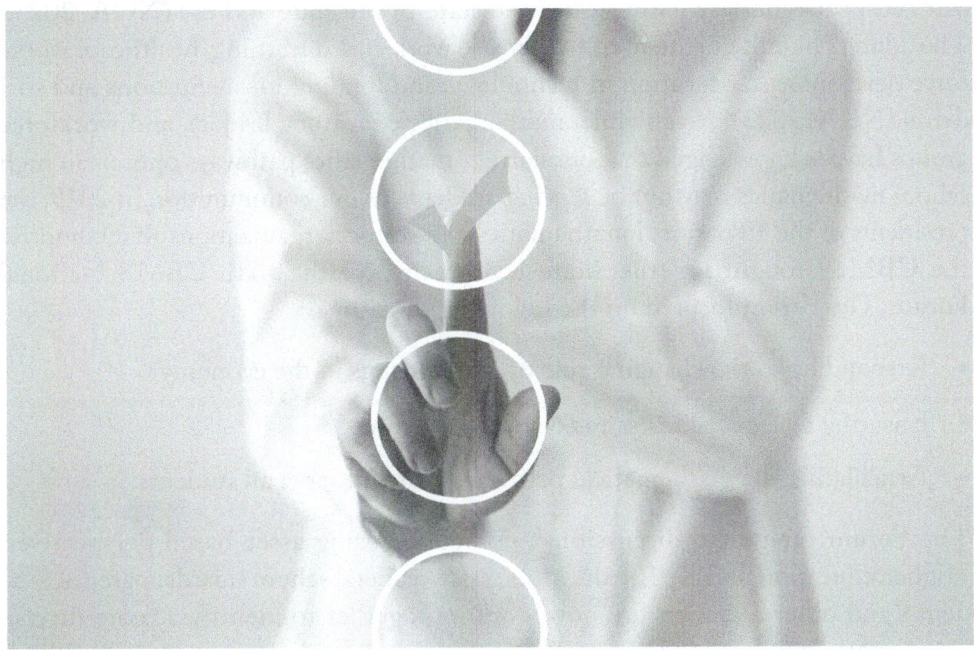

SOURCE: iStock.com/Leylaynr

Sometimes asset-based initiatives start with a need for change. The High School Pathways initiative spearheaded by the Conference Board of Mathematical Sciences (CBMS) and the Charles A. Dana Center at the University of Texas at Austin (CBMS, 2024; Charles A. Dana Center et al., 2022) is an excellent example of how states and school systems work together to assess and adjust course-taking options for students. Instead of one pathway, in which all students take courses that build toward calculus, this initiative is taking an asset-based approach to creating options for students based on their interests while aligning with the needs of our workforce. These interests may include taking calculus, but they may also lead to other rigorous course options.

Did you know that it was in 1892 that the National Education Association (NEA) created the Committee of Ten that designed our current high school structure? Among many other decisions, mathematics was designed so that students would take algebra, geometry, algebra 2, and then advanced algebra and trigonometry in high school (NEA Committee of Ten, 1894). At the time, this structure expanded the scope of public education (Holcomb, 2021), but only a small percentage of children were attending and graduating from high school at that time. As you might imagine, the focus from more than 130 years ago is deficit based. And retaining this single canonical pathway prevents us from seeing opportunities and modifying course offerings that create access to courses that are more relevant to student's goals and their lives (and the current reality of the work world).

Postsecondary education and workforce pathways become more complex as our world becomes more complex. In 2023, The National Governors Association prioritized postsecondary pathways in the state of the state address (NGA, 2023). The address highlights the need to increase work-based learning, healthcare workforce development, education in technology, and funding for institutions and students (NGA, 2023). Meanwhile, mathematics educators, leaders, and workforce groups have acknowledged that our current mathematics pathways options in high school mathematics may not be serving students or our communities. In 2019, the presidents of the 19 national mathematics and science organizations that comprise the CBMS brought together state-level teams for the sixth CBMS National Forum. The Forum focused on the following three areas:

- Responding to the changing role of mathematics in the economy

- Ensuring college readiness today and tomorrow

- Articulating the mathematical pathways that will serve all students

The Forum catapulted discussions and actions using asset-based perspectives. Higher education and K–12 educators, state leaders, school boards, parents, students, and other members are now working together to identify a more diverse set of mathematics pathways (collections of courses through middle and high school mathematics) that address the areas from the Forum and that include more than just a single sequential set of courses in middle and high school math. Several states, including Washington, Oregon, Utah, Georgia, Ohio, and Wisconsin, have already modified their pathways. This includes modifying course options so that there are more opportunities for students to take meaningful mathematics classes that align with their postsecondary goals. The collections of courses are pathways to postsecondary goals, not to be confused with additional course options that have historically led to dead-end math classes that are largely review of material from earlier grades and courses (NCTM, 2018). This effort aligns with the second bullet from the CBMS Pathways Forum: Ensuring college readiness today and tomorrow. The description includes the following quote: "High school and college mathematics teachers are recognizing the need for college-ready students, but also student-ready colleges" (CBMS, 2024). CBMS societies, including the NCTM, the Association of State Supervisors of Mathematics (ASSM), and the National Council of Supervisors of Mathematics (NCSM), acknowledge the need for a broader understanding of how mathematics is and will be used, encompassing modeling, statistics, and data science in addition to calculus (NCTM, in press). Likewise, organizations and states are addressing equity issues related to this complex situation. Issues such as barriers that may prevent certain students from accessing courses or a lack of communication or teacher preparation makes it so that some courses are more rigorous or meaningful than other courses. Take time to reflect on course offerings in your area by engaging in Try This.

What pathway options are offered in your district, school, and state? Do all students have access to these courses? What prerequisites or barriers are in place that may prevent students from taking courses based on their interests? What issues of equity exist when considering pathway options? ■

As with many worthwhile initiatives, we must unpack the structure and identify asset-based and deficit-based perspectives. Bringing in all team members and providing space for different points of view creates opportunities to enhance our system. Whether your current role is at the state, district, or school level, you need to be informed and to use your voice. Just as we want students to learn to advocate for themselves, we also must advocate for mathematics education that is best for students.

CLASSROOM WALKTHROUGHS

Classroom walkthroughs are nonevaluative, can increase student outcomes, and have the potential to build capacity of staff and increase collaboration among teachers, administrators, instructional leaders, and other team members (Rouleau & Corner, 2020). We realize several structures are similar to classroom walkthroughs, such as learning walks or classroom observations. To keep it simple, when we say *classroom walkthrough* in this section, it refers to instances in which adults are in the room for the purpose of collecting data to identify trends, not to evaluate or judge the teacher in the classroom.

Having been part of various walkthrough experiences, my (Joleigh) perspective is that asset-based walkthroughs for teachers start with what is going well or with highlighting bright spots. As a leader, if we think about what we want to see in our classrooms, we should expect the same of ourselves. Examples include highlighting the strengths and assets of individuals and the team, include data based on clear evidence, and ensure all team members have clarity about the tool being used to collect data. Clarity of the tool includes what data are being collected, clear understanding of how the data are collected, and why the particular data are important. The tool is created based on school or district goals. The Learner Centered Collaborative described that data are collected and used to spot best practices to celebrate with teachers, recognize professional learning needs, and identify aspects that contribute to or prohibit student-centered learning (Griffins, 2023).

In addition, I find that if the rubric (or other data tool) is created or revised with input from all team members including teachers, the school has a stronger common vision. Having teachers also use the tool to complete walkthroughs can provide insight for their own work and can calibrate expectations. Working together builds trust between teachers, administrators, instructional leaders, and anyone else involved.

What other characteristics are important for classroom walkthroughs?

The focus of the walkthrough is shared with teachers (better yet, co-constructed with teachers) in advance and aligns with school- and district-wide goals. Just as we start with student assets in the classroom, walkthroughs begin by noticing what the teacher does to support student learning.

■ ■ ■ Digging Deeper
Individual Education Programs

For years, data have shown that the percentage of students who reach proficiency on end-of-year state assessments decreases from one grade level to the next. Therefore, as students progress from grade to grade, fewer and fewer are meeting the standards. Although this trend occurs across all student groups, it is often a steeper decline among students with disabilities. I (Joleigh) have found these data repeatedly across districts and states and encourage you to look at your own local or state data to examine trends in your area. If this is our reality, how do we change this trajectory? What do we modify so that our students with disabilities are successful with grade/course-level content? ■

As we have discussed earlier, according to IDEA, students who are identified as having a disability, including those who are neurodiverse, have a right to an IEP. The term *Individualized Education Program* or *IEP* means a written statement for each child with a disability that is developed, reviewed, and revised at least once per year by all IEP team members. The team members include the parents, teachers, school administrators, related services personnel, and the student (as appropriate). The IEP includes PLAAFP that begins by using data (quantitative and qualitative) to identify student strengths. In addition to strengths, areas of need to support progress in the general education curriculum are identified. IEP goals are written to align with the PLAAFP and include Specially Designed Instruction (SDI), Least Restrictive Environment (LRE) to the maximum extent appropriate, and other factors to ensure students have access to and success with grade/course-level content.

What does this mean for asset-based learning environments? It means that we must look at current practices and identify what we are already doing that is asset based and consider tweaks to enhance our work. More importantly, it means we must name deficit-based practices and disrupt what does not work. One place to start is by looking at how our IEPs are written. How do they support the intention of IDEA from an asset-based perspective? In what ways do our current IEPs use deficit-based language that may prohibit students from accessing and having success with grade-level content?

I (Joleigh) began this work more than ten years ago and have been amazed at the results of those committed to transforming IEPs in their classrooms, schools, and districts. In 2020, I was asked to oversee and co-create the IEP Reflective Framework for the Utah State Board of Education (USBE, 2024). This framework supplies guidance and reflective questions for all IEP team members as it intertwines the relationship between compliance and best practices. Using this framework, four district leaders and I set up a strong collaborative team among general education and special education teachers that would disrupt deficit-based decisions in how we had been creating IEPs and embrace an asset-based stance on what students can do. The shared vision of teachers, administrators, parents, and the leadership team resulted in extraordinary results for students in the first year of implementation during the 2022–2023 school year. The team worked together with the following goals driving our work:

- **Improve student outcomes,** resulting in confident and successful students

- **Increase student access to and success with grade-level content,** using the major work of the grade to support students learning meaningful content and identifying individual student assets and strengths

- **Empower students by providing choice,** increasing student and teacher content knowledge of explicit strategy instruction with a focus on developing a positive student mathematics identity

In just one year, all school-level teams across the four districts saw a significant increase in student achievement. As many schools typically see a decrease in proficiency from one grade to the next, all teachers who participated saw their students with needs data increase. In addition to proficiency, Utah uses a median growth percentile (MGP) measurement. Thus, students who scored similarly the previous year are compared with each other based on their growth or their scores for the current year. All district schools that took part saw an increase in this area as well, with some noticing that even though our focus was on supporting students with needs (as identified in state-level disaggregated data), that students in both general education and special education showed more growth than did their peers in other schools across the state. The teachers indicated that this was a direct result of the shift in how goals were created using the resource documents provided, as well as a result of the collaboration between general education and special education teachers. As one teacher said, "Our focus was to increase outcomes for students with needs, but all of our students benefited as a result."

Data from each district can be seen in Figures 9.1 through 9.4. What do you notice about the data?

FIGURE 9.1   Data From District A, School 1

School District A, School 1
Fifth Grade Students with Disabilities

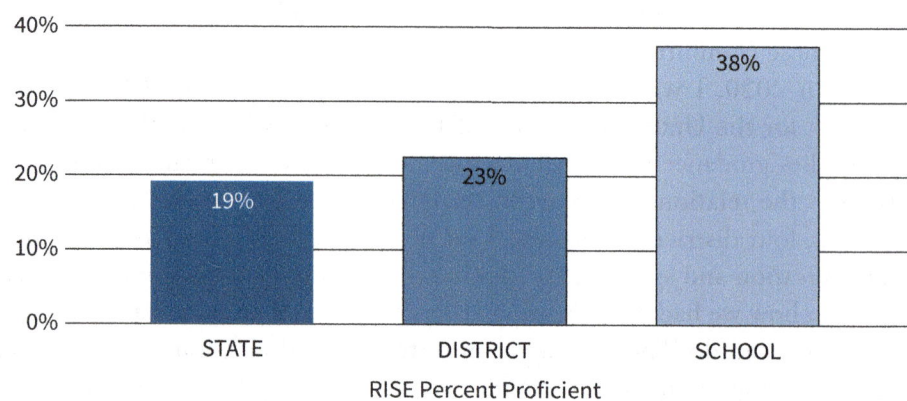

FIGURE 9.2   Data From District B, School 1

School District B, School 1
Fourth Grade All Students

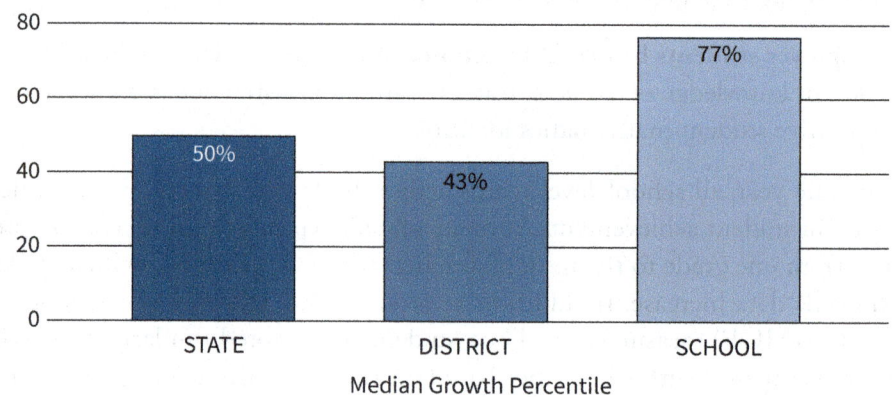

FIGURE 9.3   Data From District C, School 1

School District C, School 1
Fourth Grade All Students

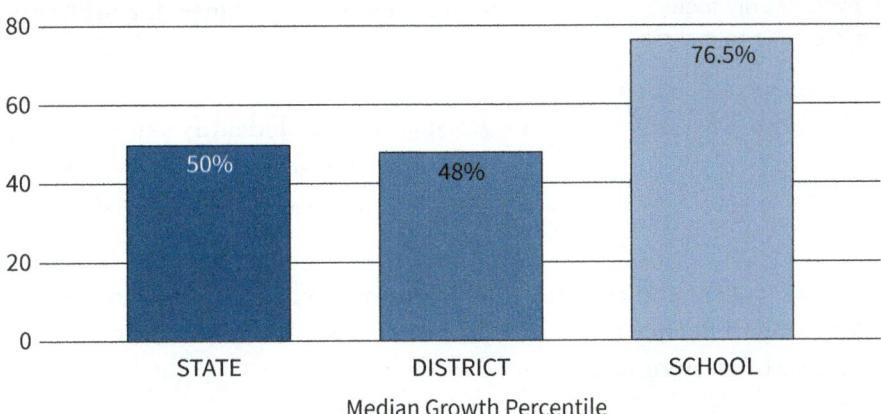

Median Growth Percentile

FIGURE 9.4   Data From District D, Schools 1 and 2

School District D, Schools 1 and 2
3rd Grade Operations and Algebraic Thinking

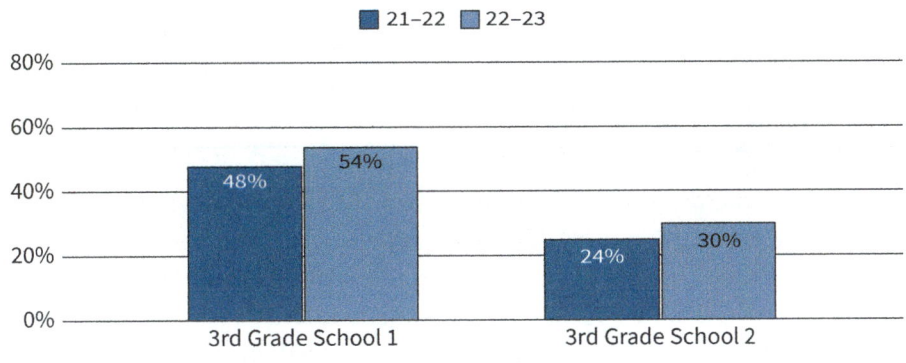

RISE Percent Proficient

In addition to quantitative data, students and teachers shared many positive outcomes from our first year of this work. The following two quotes from participating teacher reflections are captured here:

- "As a result of this project, I feel like we all have a lens of looking for strengths in student work."

- "I believe the biggest difference is that we believe our students can reach high expectations and that re-looking at how we teach benefits all of our students."

Tip

Do an audit of your IEPs. If they are generic (solving word problems), below grade level, or only focus on SMPs, then it is time to make a change.

We highly encourage including SMPs in goals, but they are an equity trap if they are not also connected to grade-level content. When making the change to include using asset-based perspectives to increase outcomes for students with needs:

- Review IEPs, and identify student strengths and content so that all students with needs have access and success with grade level content.

- Include student choice in how they show and solve problems, and use the definition of fluency (flexible, efficient, accurate) to ensure students gain mastery of meaningful content.

- Promote a positive mathematics identity for students by positioning them as capable contributors to important mathematics.

- Incorporate time for the special education and general education teacher to collaborate, look at student work, show student strengths, celebrate success, and deepen our own understanding of strategies.

- Use resources to support team members in making the shift.

Reflect, Apply, Transform

Return to the structures in the Alignment exercise and select one to focus on (or propose a new one of your own). Choose a structure or area about which you feel the most passionate. Unpack the intention of this structure, and revisit asset-based and deficit-based elements of its current implementation. Identify components that you would like to address. What can be enhanced, or what may need more attention? Is making changes in this structure something in your purview?

Next, identify the constituencies that have a voice in this structure: students, mathematics teachers, school-based coaches, district-based personnel, regional personnel, state-level personnel, families, and community members. Unpack the intention of the structure. We need to understand how decisions have been made before moving forward with discussions about improvements or changes.

This exercise is designed to help you identify the pieces of the structure that are valuable and worth retaining, as they reflect asset-based perspectives, and the aspects of the structure that are worth taking a deeper look. Are there aspects within our control that could be enhanced? Are there systems to unpack and change? This exercise also helps you identify the allies you might recruit to support change for this structure within the system. Chapter 10 will identify some potential action steps that you can take to enact change.

Now let's revisit the following questions:

1. What aspects of the system cultivate asset-based perspectives?

2. What aspects of the system don't support or actively work against my classroom efforts to foster asset-based perspectives?

3. Where do I have agency to encourage change to my systemic structures?

CHAPTER 10

Leveraging Strengths Across Our Systems to Advocate for Long-Term Change

As we've seen in the past two chapters, systemic structures involve many people with many different roles. It can be challenging to figure out where you might have leverage, how to build momentum, where to find allies, and how to identify and overcome challenges and obstacles to support systemic change. In addition, this sort of change requires sustained long-term efforts. It can be daunting!

We close the book with a discussion of how to advocate for systemic change. We'll share some practical tools and strategies that you can use in your role to get conversations started about asset-based systems and to specifically surface challenging deficit-focused ideas for discussion, consideration, and ultimately revision. This chapter will read a bit differently than some of the others with the focus on specific tools and supports.

> *"The surest way to limit one's impact is to attend to only one piece of a system, or to only one of these systems, without regard for how it affects the other pieces and systems." (National Council of Supervisors of Mathematics [NCSM], 2014, p. 8)*

Questions to Consider

- Where are the points of entry I have into conversations about fostering asset-based systems?

- What strategies can I use to foster challenging conversations about facets of our systems that reflect deficit-based perspectives?

- How can I develop my vision for identifying challenges to change within the system and working proactively to address those challenges?

Both of us have had the opportunity to lead one of our major mathematics education professional organizations, and we are currently serving on the board of the National Council of Teachers of Mathematics (NCTM) together as we write this book. I (Mike) remember that one of the things that attracted me to this sort of professional service was the desire to think big and to support our professional organizations as they grew and changed to support their members. I know that early on when I first started serving on organizational boards and committees, I entered with an ambitious set of ideas that I wanted the organization to consider. For example, I came in wanting to see the organization offer more direct professional learning opportunities, to diversify its membership in particular ways, and to think about changes to the structure of the annual meeting that would be more interactive and engaging.

What I've found after serving a few different organizations is that there's a lot we don't understand from the outside about how the systems are constituted and what is entailed in making changes to those systems. Sometimes when people start to see all the interconnected aspects of the system and hear about why things are done the way they are done, it can be demotivating. This story can be heard (and, honestly, told as well) as reasoning for why nothing should change, that the group has already thought through a similar idea in the past. But rarely have I found that this story is intended to be heard in this way. Rather, it's an effort for members of the group's leadership to illuminate the complexity of the system.

I'll admit early on that I found these stories to be daunting and tiring, with my reaction often being to back off of my ambitious idea for change. But over time, I came to understand that seeing the complexity of the system allowed me to revise and fine-tune my idea to actually be more achievable by fitting it more soundly into the structures and systems of the organization. This understanding also helped attune me to the need to ask good questions and understand the operating of the system to help identify which of my (usually too numerous) ambitious ideas for change might make sense to begin with and which might likely to take more time and energy to bring to fruition.

Our systems are never perfect. You probably entered this book with some existing frustrations about aspects of your systems that you would like to see change. In reading through the last few chapters of this book, you may have noticed even more areas that are ripe for change. This can be frustrating! Sometimes when we think about change to our systems, we see two options: having to compromise our ideals and values to work within the system or a desire to oppose aspects of the system in order to change it. We understand this feeling! We'd like to challenge you to think this contrast less as a dichotomy and more as a complementary set of approaches. There are times when working inside our existing systems strategically is a way to effect change. There are also times when making principled objections to aspects of a system is the most effective change-making strategy. Just as we've thought about a continuum of deficit-to-asset perspectives, we encourage you here to consider a continuum of actions and interactions in which to engage as we act as agents of change.

ALIGNMENT EXERCISE: ASSET-BASED PERSPECTIVES AND SCHOOL STRUCTURES

In Chapters 8 and 9, we identified several systemic structures that impact mathematics teaching and learning that may or may not reflect asset-based perspectives, including:

- Professional learning communities (PLCs)

- Grades and grading systems

- How students are placed in classes or courses

- How teachers are placed in classes or courses

- How teachers are evaluated

- How we recognize student performance in mathematics with the community

- How we talk about mathematics with the community (including other adults in our school or district)

Let's take a look at one of these structures in depth.

■ ■ ■ Try This

Take one of the structures from the list above that you feel, in your current context, has room to develop in more of an asset-based way. Using that structure, imagine a conversation with your colleagues, community, and/or students that directly addresses the concerns. You might document this conversation in various ways: as an imagined dialogue between team members, in a storyboard that describes the key points that you would hope would be addressed, or as an acted-out audio dialogue. After documenting the potential conversation, reflect on what you see as the key points of action that would come from such a conversation that would shift the structure to being more asset based. ■

Writing an imagined dialogue or storyboard can be challenging. It may not come naturally to you or feel authentic—that's okay! The important idea behind this exercise is that it allows us to take the perspective of a variety of other colleagues as we write down what we think they might say or do. We provide a short example of an imagined dialogue among how students are placed in classes or courses below.

The scene:	It's time to start building next year's course schedule, and Mr. Gonzalez has been reading quite a bit about the detrimental impacts of tracking. He feels strongly that the Glen Valley High School math department needs to start moving in this direction. He plans to bring this topic up in their department meeting. Mrs. Nicolet, the department chair, has taught a course load of all Honors and Advanced Placement (AP) courses for years and expects that she might see this suggestion as challenging her courses directly. Ms. Elias, a second-year teacher, has talked with Mr. Gonzalez and agrees with his stance, although she worries about whether she has the standing to speak up in the faculty meeting. Mr. Francis has historically requested to teach a course schedule consisting of both Honors courses and the track for students with a history of mathematical struggles, and Mr. Gonzalez wonders if this will make him an ally in this conversation.
Mrs. Nicolet:	Okay, let's get started. I've begun by copying over last year's schedule into a new spreadsheet for this year. We can use this as a starting point, but I want to make sure we all understand that we can make any changes we'd like. Take a minute to look things over, and let's talk about what we might want to do together.
	Oh—and I know in the past we'd sent in preference sheets about what we want to teach. You might notice we didn't do that this year. I'm hoping we can make the process a little more transparent and talk about our preferences with each other. I know that may be a little different, but let's give it a try, okay?
Mr. Francis:	You know how I like to put my schedule together and we've got that down, so I'm good. If anyone else might be interested in Honors Algebra II though, it's been a while since I've taught Honors Precalculus and I might be willing to trade.
Mrs. Nicolet:	I was thinking that Ms. Elias has had a tough schedule this year and maybe she's earned herself an Honors course. Ms. E, what do you think about Honors Algebra II?
Ms. Elias:	*pauses* Well um, I guess that would be okay.
Mr. Gonzalez:	I was wondering if we could talk about the Honors courses overall. I know there's been some frustration about figuring out placements and we've been trying to be more generous about letting students in who wanted to take the course. But some of the inequities have been really bothering me. I think we've all had students who weren't placed into that class at the start of the year that by

	the end we know could have been successful. I was hoping we could talk about a different path forward—maybe not having separate Honors sections.
Mr. Francis:	I'm sending all those angry parent emails to *you*, Mr. G.
Mrs. Nicolet:	*chuckles* Okay, that probably wasn't strictly necessary, but there's some truth behind that. If we eliminate Honors, we will have a revolt on our hands.
Mr. Gonzalez:	I couldn't agree more. But I don't think that's what we should do—well, not exactly. I've been emailing with a few teachers in some other districts who have tried models in which everyone has the *opportunity* to earn an Honors designation but everyone is in the same section. There are a few different ways to do it—pros and cons of course, but it really resonates with me in terms of equity for our students.
Ms. Elias:	Yeah. I had a lot of students in the regular Algebra I sections this year that I think would've been ready for more. I wish I'd had the time to challenge them more. Do you have models for us to look at?
Mr. Gonzalez:	I could get us some. I was interested though before we dive into the models about how people feel about the general idea. This would be more work and a different sort of work, and we'd have to think about how to talk to parents about what we're doing and why. And we've got to convince admin that it's a good idea. Can we open a dialogue about how a move like this would, or would not, reflect our math department mission statement?

Thinking about how difficult discussions will go has some significant benefits. Just like when we anticipate student thinking for a lesson we're going to teach, thinking through the comments people might make in such a discussion in advance helps us think about how to respond to them in ways that are asset focused, that connect to the core values of our department and school, and that seek to invite others into a collaborative dialogue. We're going to explore a few more examples and think about specific strategies and tools that are likely to support constructive asset-based discussions of systemic structures.

HOW CAN I INITIATE PRODUCTIVE CONVERSATIONS ABOUT MOVING TOWARD MORE ASSET-BASED SYSTEMS?

Let's unpack four examples of conversations that might be particularly important for moving toward more asset-based systems. We selected these also because they can be particularly challenging discussions, as people are likely to have strong feelings. A few important common tenets you can use across all these conversations:

- We are all on the same team: we are interested in supporting stronger student learning outcomes.

Whether we're talking to colleagues, administrators, or parents, we all have the learning and well-being of our students at heart. It's why we are in education and why we work so hard to support students in building on the assets they bring to the classroom.

- We are all acting on what we believe to be best for students.

Even when we are advocating for very different instructional practices, we are doing it from a place of believing that it is best for students. One goal might be to surface those beliefs and ultimately shift and change unproductive beliefs, but we must first acknowledge that people are operating from a place of authenticity, not trying to undermine one another or just being contrarian.

- We can approach difficult conversations with genuine curiosity and seeking to understand one another's ideas.

Fostering change is hard—you're asking people to do something that they are not currently doing. Opening up space for people to be curious and seeking to understand is an important prerequisite for conversations that truly promote systemic change.

Let's look at how these principles operate in the context of conversations.

SURFACING VALUES ABOUT STUDENT LEARNING WITH YOUR PLC

PLC meetings sometimes skim the surface of important ideas, with time taken up handling paperwork and logistics. Moving PLC discussions toward deeper discussions of teaching and learning requires discipline. Once a PLC begins moving in that direction, opportunities arise to dig into underlying beliefs about student learning. Teachers often have underlying and deeply held beliefs about how students learn and what students can learn that may reflect deficit perspectives. PLC discussions about planning lessons and assessing student thinking can provide an asset-based setting to thoughtfully examine those beliefs. Two key activities can frame these discussions: planning lessons together and analyzing student work.

Planning Lessons Together

Planning lessons together can be very powerful, but this activity can also focus more on the things we will do as teachers versus on the thinking and learning of students. Balancing attention to teacher moves and student thinking is important during these discussions. Several questions can be used to help deepen the collaborative planning of lessons:

- What are the most important ideas that students are likely to bring to this lesson when it starts?

- What are the key ideas we want students to take away from this learning experience?

- What are the elements of that learning that we expect to hear and see as the lesson unfolds?

SOURCE: iStock.com/SDI Productions

As teachers discuss what ideas students might bring and how learning develops, opportunities often present themselves to probe why teachers might think certain ideas are important building blocks for learning. For example, teachers sometimes overestimate the knowledge that they think students need to be successful in a lesson. Asking questions about why an idea is important for success can help to surface underlying beliefs about what students should know and who has opportunities to learn. The focus on the mathematics content also decenters students' performances, allowing teachers to deeply consider what students should currently bring to the classroom.

During these conversations, listen for qualifiers like "some students" or labels that describe students' previous performances or demographic characteristics. If teachers, for example, say that some students won't remember a key idea that they feel is necessary for success in the lesson, ask about how we might support that group of

TRANSFORM YOUR MATH CLASS USING ASSET-BASED TEACHING FOR GRADES 6-12

students and invite them into the mathematics learning. Focus the conversations on what assets students will bring to the table rather than what they might not bring with them on any given day.

Analyzing Student Work

Bringing student work for teachers to share and analyze together can be a powerful activity. It can also serve to sidestep and ultimately dismantle biases about students' prior achievement and other nonacademic factors. After planning a lesson that will be taught by multiple teachers, encourage teachers to bring student work artifacts that can be shared without student names. Discussions of the student work should focus on the key ideas teachers noted during planning that were indicators of success for the lesson. Where do we see those ideas in the student work fully expressed? Where do we see elements developing? How do these analyses help us think about planning and teaching the lesson in the future, or what we should plan and teach tomorrow?

SOURCE: iStock.com/fizkes

INTERROGATING THE ROLE OF ASSESSMENTS

Assessments take up a great deal of our time and energy in the teaching of mathematics. Structurally, the work of assessment often gravitates toward deficit perspectives. State benchmarks press us to describe what percentage of our students is below grade level or some other expectation. Summative assessments are often highly consequential for the mathematics that students have access to (or are not allowed access to) in the next grade level or course. How might we think carefully about the role of assessments?

At the classroom and school level, we encourage PLCs to discuss the role of formative and summative assessment. According to Wiggins (1993) and Wiliam (2003), formative assessment should provide students with two things: feedback on their current performance and a clear pathway to improve that performance with effort. Without the second component, formative assessments are much more likely to fall on the deficit side of the deficit-to-asset continuum. And in considering the nature of the feedback, to what extent does feedback we provide students reflect an asset-based perspective? To do so, such feedback would need to be clearly grounded in the strengths of the work that students produced on the assessment rather than solely focused on the aspects of the problem that were incomplete or incorrect.

■ ■ ■ Try This

Take the formative assessments for an upcoming unit and think in advance about how the questions asked provide students with opportunities to demonstrate their mathematical assets related to the goals of the unit. Then, for each item, discuss the sort of feedback you might give based on anticipated student performances on each item. Consider together how the feedback meets the two formative criteria: feedback on current performance and pathway to improve performance with effort. Discuss how to provide that feedback in asset-based ways. ■

Beyond the classroom and school level, state and national assessments often have an outsized influence in our work in the classroom. There are many reasons why this is the case, and although we all are likely to have frustrations with how large-scale assessments are designed, it's unlikely we have agency to change the assessments. But where we can have influence is how the assessments are perceived in the school and district. One way of changing that narrative is to collect some data from the classroom and school assessments from which you have control and compare those outcomes with the large-scale assessments.

■ ■ ■ Try This

Collect assessment information from the classroom or school level that represents an asset-based approach. These could be aggregated scores on formative assessments, summative assessments, portfolios, other tools, or a combination of multiple data sources. At the student or classroom level, correlate those outcomes with assessment outcomes from a large-scale assessment. How well correlated are they? What do you notice about the correlations? ■

As noted in Steele and Huhn (2018), a district with which I (Mike) had worked extensively performed a similar analysis to attempt to shift the district administration's focus on the ACT as an important learning outcome. They aggregated data from carefully designed assessments and end-of-course grades and correlated those with ACT scores. The result was a low correlation. The district then used this to frame an argument for administration: If you believe in what we do in the classroom and the asset-based ways we assess student performance, but the ACT outcomes don't line up well with how we assess student performance, what does that tell us about the ACT and what it might truly assess?

INITIATING A DEMOGRAPHIC ANALYSIS

In asset-focused mathematics teaching and learning, achievement and mathematical progress should be primarily governed by the nature of the teaching and learning in classrooms. But we know that historically, some of our educational systems (and specifically our mathematics education systems) are built in ways that systemically disadvantage learners of color and promote deficit discourses related to students in minoritized groups (Adiredja & Louie, 2020; Martin, 2019). These systemic biases can be challenging to identify and even more challenging to confront and change. One activity that can help with this challenge is to conduct regular demographic analyses of course enrollment and/or achievement in secondary mathematics classes.

■ ■ ■ Try This

Collect enrollment and/or achievement data from the courses offered in middle and high school, and break enrollment and achievement data down by demographic categories important to you and your school or district. This breakdown might include race, gender, special education status, socioeconomic status, or geographic community or neighborhood. These demographic categories should not predict course enrollment or achievement.

Discuss with your colleagues what you notice and what you wonder about. Then take follow-up action. If you notice groups that are systematically disadvantaged, interview students in individually and in small groups about their mathematics learning experiences. Determine where the barriers are to more equitable representation and make a plan to dismantle them. ■

Demographic analyses can be very simple to do—the central office of most schools and districts have these data easily available. Building a simple comparative table does not require a great deal of time or energy. Focusing on noticing and wondering about those data can help to keep the conversation focused on the current realities of the situation rather than on trying to explain away inequities.

MESSAGING TO THE COMMUNITY

Some of the most persistent deficit-based messages about mathematics reside in our community. We have all heard people taking pride in being bad at mathematics, attributing mathematical performance to genetics, and furthering the perception that mathematics is a challenging activity that makes little sense to most and is only done at a high level by a preordained few. This perception in and of itself is problematic, but it can also cause significant challenges when making changes to a mathematics program. Specifically, families of students who are successful at mathematics may be resistant to changes, seeing it as a threat to their student's current or future status. Families with students that have been successful in deficit-based systems may not see a need to change, particularly if the parents themselves were successful in such a system. Messaging to the community about what an asset-based approach to mathematics looks and sounds like is critical to the success of any change initiative in the teaching and learning of mathematics.

SOURCE: iStock.com/SDI Productions

■■■ Try This

Prior to making a new change to your mathematics program, hold a listening session for families about mathematics. Discuss the mathematical assets that they use in their professional and personal lives. Model doing mathematics together using tasks with low barriers to entry and multiple pathways and highlight the important mathematical practices and process in which they are engaging as they do the tasks. Use this instructional interaction as a springboard for discussing mathematical assets and leveraging what students bring to their classrooms. ■

Building community support for change can be challenging. As soon as possible, we encourage you to begin working with your community and examining the extent to which the community holds deficit-based perspectives about mathematics teaching and learning. Work to model more asset-based approaches both before, during, and after systemic changes to the mathematics program.

Reflect, Apply, Transform

This section has provided a few detailed examples of challenging systemic structures and ways to begin conversations to move them along the asset-based perspectives continuum. Your starting point may be similar to one of these examples, or it may be different.

First, revisit the questions you considered at the start of the chapter. How has your thinking shifted about your points of entry, the strategies you have for having challenging conversations, and how you can develop your vision for a more asset-based system?

- Where are the points of entry I have into conversations about fostering asset-based systems?

- What strategies can I use to foster challenging conversations about facets of our systems that reflect deficit-based perspectives?

- How can I develop my vision for identifying challenges to change within the system and working proactively to address those challenges?

Next, reflect on one of the structures you identified in your Alignment exercise in chapters 8 and 9. Identify how the current structure is situated on the deficit-to-asset continuum. Determine where you have agency in making change and where others have agency.

Then, plan a first conversation with a small group (departmental colleagues, school leadership, community) to begin examining the system and identifying what aspects of the system could be more strongly aligned with an asset-based perspective. Use or adapt the action steps discussed in the scenarios in this chapter.

What Now?

Thanks for taking the journey with us in thinking about how we can transform our classrooms to reflect more asset-based perspectives in our language, routines, and systems. Just as deficit- and asset-based perspectives are part of a continuum, so too is the evolution of our classroom practices. We hope that your experience with this book represents our own in writing it—at the start, we were cognizant of the need for stronger asset-based perspectives. As we wrote and reflected on our practice as mathematics teachers and mathematics teacher educators, we were able to identify countless facets of our practice that were more deficit based than we may have liked. We had meaningful discussions between ourselves and with our colleagues about new insights into our language, routines, and systems. We reflected on past practices that we had not realized were closer to the deficit side of the continuum and considered ways that we might shift toward the asset side.

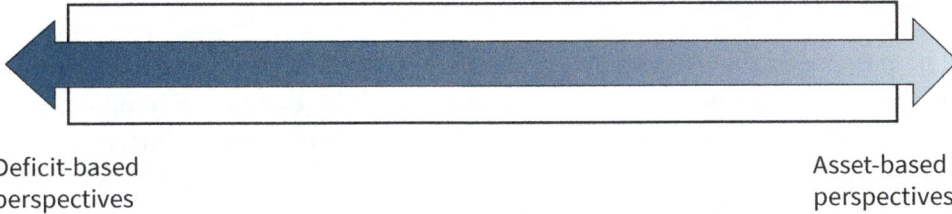

Deficit-based
perspectives

Asset-based
perspectives

We imagine your journey might be similar. So what now? We hope you've come to the end of this book with more ideas about what you might want to change than you entered with. We also hope that you've come away with some concrete and actionable ideas to try. This might feel a bit overwhelming. We encourage you to spend some time reflecting on the things you'd like to change and on identifying actions that are short-, medium-, and long-term ambitions for change. In our experience, it's critically important to have all three as it allows us to get some quick wins with things we can change immediately, plan for some areas of work in the next year or handful of years, and set some more ambitious goals related to bigger or more systemic change to work toward. We also hope that if you haven't already, you share this work with colleagues and recruit them as allies, critical friends, and co-conspirators in the journey toward change. What are the assets that they can bring to the work to make the collective effort stronger?

PERSONAL STORY

We have shared that we started this journey to improve our own work. One of the many aspects we have noticed by digging into this work is that the deficit-to-asset continuum has gone beyond experiences in our professional lives and has moved into our personal lives. Can you imagine? We spend less time worrying about trivial matters as we have grown to spend more time focusing on valuing the assets of others. For example, my (Joleigh) daughter has her own business and works many hours each week. I have shifted from a deficit perspective of always worrying about her lack of experience as a business owner to now focus on her strengths as a designer, collaborator, and project manager. I now hear and see her as a strong, capable business woman, and my conversations with her are much more empowering. I now find myself constantly thinking, "How do I listen for strengths and hear the contributions of others in my everyday conversations?" Imagine if all of our interactions had us thinking, "How am I increasing the sphere of belonging? What actions can I make to ensure those around me see themselves as capable?"

ALIGNMENT EXERCISE: REFLECT, APPLY, TRANSFORM

The purpose of this Alignment exercise is to celebrate you, your progress in this journey, and those you support. This exercise can be done after completing the book, or it can happen anywhere throughout the book (after a part, chapter, or even section of the book). Begin by selecting something that stood out for you from the reading and the subsequent reflections and actions made. Answer the following questions:

1. Reflect: What about this topic(s) is important to you?

2. Apply: What shifts have you made as a result?

3. Transform: How have these shifts transformed your practice? How has this transformation increased the sphere of belonging? Describe your evidence that those you support see themselves as capable? How have your beliefs shifted about what is possible (or what can be done by others)?

By now, we hope you have experienced that our intention is not a book that you pick up, read, and move on to the "next thing." Our profession is complex and rewarding, especially when we reflect on our own practices, apply new understandings, and transform our classrooms using asset-based teaching. Because when we know better, we do better.

References

CHAPTER 1

Aguirre, J. M., Mayfield-Ingram, K., & Martin, D. B. (2024). *The impact of identity in K-12 mathematics: Rethinking equity-based practices* (2nd ed.). National Council of Teachers of Mathematics.

Association of State Supervisors of Mathematics & Association of Mathematics Teacher Educators. (2024). *Position statement: Asset-Based Perspectives in Mathematics Education.*

Boaler, J. (2022). *Mathematical mindsets: Unleashing students' potential through creative mathematics, inspiring messages and innovative teaching* (2nd ed.). Jossey-Bass.

CAST. (2018). *Universal Design for Learning guidelines version 2.2.* http://udlguidelines.cast.org

Collaborative for Academic, Social, and Emotional Learning (CASEL). (2024). CASEL Framework. https://casel.org/fundamentals-of-sel/what-is-the-casel-framework/#interactive-casel-wheel

Dweck, C. (2016). What having a "growth mindset" actually means. *Harvard Business Review, 13*(2), 2–5.

Individuals with Disabilities Education Act, 20 U.S.C. § 1400 (1990; reauthorized 2004). https://sites.ed.gov/idea/regs/b/d/300.320

Kelemanik, G., Lucenta, A., & Creighton, S. J. (2016). *Routines for reasoning: Fostering the mathematical practices in all students.* Heinemann.

Kobett, B. M., & Karp, K. S. (2020). *Strengths-based teaching and learning in mathematics: Five teaching turnarounds for grades K-6.* Corwin.

Lambert, R. (2021). The magic is in the margins: Universal Design for Learning

Math (UDL Math). *Mathematics Teacher: Learning and Teaching Pre-K–12, 114*(9), 660–669.

Lambert, R. (2024). *Rethinking mathematics and disability: A UDL Math classroom guide for grades K-8.* Corwin.

National Council of Teachers of Mathematics (NCTM). (2014). *Principles to actions: Ensuring mathematical success for all.*

National Council of Teachers of Mathematics (NCTM). (2018). *Catalyzing change in high school mathematics: Initiating critical conversations.*

O'Keefe, P. A., Dweck, C. S., & Walton, G. M. (2018). Having a growth mindset makes it easier to develop new interests. *Harvard Business Review.* https://hbr.org/2018/09/having-a-growth-mindset-makes-it-easier-to-develop-new-interests

Terada, Y. (2020). *Classroom management mistakes—and the research on how to fix them.* Center of Excellence to Prepare Teachers of Children of Poverty. https://www.fmucenterofexcellence.org/bestpractice/7-classroom-management-mistakes-and-the-research-on-how-to-fix-them/

CHAPTER 2

Gibbons, P. (2009). *English learners, academic literacy, and thinking: Learning in the challenge zone.* Heinemann.

Halliday, M. A. K., & Webster, J. J. (2003). *On language and linguistics: Volume 3.* A&C Black.

Herbel-Eisenmann, B. A., Cirillo, M., Steele, M. D., Otten, S., & Johnson, K. (2017).

Mathematics Discourse in Secondary Classrooms (MDISC): A case-based professional development curriculum. Math Solutions.

Jansen, A. (2020). *Rough draft math: Revising to learn.* Stenhouse.

Lambert, R. (2021). The magic is in the margins: Universal Design for Learning Math (UDL Math). *Mathematics Teacher: Learning and Teaching Pre-K–12, 114*(9), 660–669.

Lambert, R. (2024). *Rethinking mathematics and disability: A UDL Math classroom guide for grades K-8.* Corwin.

Liljedahl, P. (2021). *Building thinking classrooms in mathematics, grades K-12: 14 teaching practices for enhancing learning.* Corwin Mathematics.

Open Up Resources. (2021). *Open Up resources high school mathematics integrated math I.*

Ruef, J. L. (2020). What gets checked at the door? Embracing students' complex mathematical identities. *Journal of Humanistic Mathematics, 10*(1). https://doi.org/10.5642/jhummath.202001.04

CHAPTER 3

Bennett, C. A. (2014). Creating cultures of participation to promote mathematical discourse. *Middle School Journal, 46*(2), 20–25. https://doi.org/10.1080/00940771.2014.11461906

Jansen, A. (2020). *Rough draft math: Revising to learn.* Stenhouse.

Lambert, R. (2024). *Rethinking mathematics and disability: A UDL Math classroom guide for grades K-8.* Corwin.

SanGiovanni, J. J., Katt, S., & Dykema, K. J. (2020). *Productive math struggle: A 6-point action plan for fostering perseverance.* Corwin Mathematics.

Tannen, D. (2021, September 25). In real life, not all interruptions are rude. *The New York Times.* https://www.nytimes.com/2021/09/25/opinion/interrupting-cooperative-overlapping.html

CHAPTER 4

Individuals with Disabilities Education Act, 20 U.S.C. § 1400 (1990; reauthorized 2004). https://sites.ed.gov/idea/regs/b/d/300.320

Johnson, K. R., & Fonbuena, L. C. (2023). Positionalities in our practices and papers. *Mathematics Teacher Educator, 11*(3), 145–154.

Kitchen, R. S., Anderson Ridder, A., & Bolz, J. (2016). The legacy continues: "The Test" and denying access to a challenging mathematics education for historically marginalized students. *Journal of Mathematics Education at Teachers College, 7*(1), 17–26.

Kitchen, R. S., & Berk, S. (2016). Educational technology: An equity challenge to the Common Core. *Journal for Research in Mathematics Education, 47*(1), 3–16.

Kitchen, R. S., Garcia-Olp, M., & Van Ooyik, J. (2017). Resisting student labeling in this era of testing. *Colorado Mathematics Teacher, 50*(3), 1.

National Council of Teachers of Mathematics (NCTM). (2018). *Catalyzing change in high school mathematics: Initiating critical conversations.*

National Council of Teachers of Mathematics (NCTM). (2020). *Catalyzing change in middle school mathematics: Initiating critical conversations.*

Rehabilitation Act, Pub. L. 93–112, 87 Stat. 355 (1973).

Steele, C. M. (1997). A threat in the air: How stereotypes shape intellectual identity and performance. *American Psychologist, 52*(6), 613–629. https://doi.org/10.1037/0003-066X.52.6.613v

CHAPTER 5

Baylor University. (2024). Formative assessment. Academy for Teaching and Learning. https://atl.web.baylor.edu/guides/assessing-student-learning-and-teaching/formative-assessment

Brown, P. C., Roediger III, H. L., & McDaniel, M. A. (2014). *Make it stick: The science of successful learning.* Harvard University Press.

Collaborative for Academic, Social, and Emotional Learning (CASEL). (2024). CASEL Framework. https://casel.org/fundamentals-of-sel/what-is-the-casel-framework/#interactive-casel-wheel

Frey, N., Fisher, D., & Smith, D. (2019). *All learning is social and emotional: Helping students develop essential skills for the classroom and beyond.* ASCD.

Hattie, J. (2008). *Visible learning: A synthesis of over 800 meta-analyses relating to achievement.* Routledge.

Hendrickson, S., Hilton, S. C., & Bahr, D. (2008). The Comprehensive Mathematics Framework (CMI): A new lens for examining teaching and learning in the mathematics classroom. *Utah's Mathematics Teacher, 1*(1), 44–52.

Hunt, J. H., & Ainslie, J. (2021). *Designing effective math interventions: An educator's guide to learner-driven instruction.* WordPress.

Leinhardt, G. (2001). Instructional explanations: A commonplace for teaching and location for contrast. In V. Richardson (Ed.), *Handbook of research on teaching* (4th ed.). American Educational Research Association.

Leinhardt, G., & Steele, M. D. (2005). Seeing the complexity of standing to the side: Instructional dialogues. *Cognition and Instruction, 23*(1), 87–163.

Lemon, T., & Hendrickson, S. (2023). Building coherence and progression on sound frameworks. *Mathematics Teacher: Learning and Teaching PK–12, 116*(7), 490–502.

Liljedahl, P. (2021). *Building thinking classrooms in mathematics, grades K–12: 14 teaching practices for enhancing learning.* Corwin.

National Council of Teachers of Mathematics (NCTM). (2016). *Position statement: Providing opportunities for students with exceptional promise.* https://www.nctm.org/Standards-and-Positions/Position-Statements/Providing-Opportunities-for-Students-with-Exceptional-Promise/

National Council of Teachers of Mathematics (NCTM). (2023). *Position statement: Procedural fluency in mathematics.* https://www.nctm.org/Standards-and-Positions/Position-Statements/Procedural-Fluency-in-Mathematics/

Otten, S., Cirillo, M., & Herbel-Eisenmann, B. A. (2015). Making the most of going over homework. *Mathematics Teaching in the Middle School, 21*(2), 98–105.

Ramirez, N., & Celedon-Pattichis, S. (2012). *Beyond good teaching: Advancing mathematics education for ELLs.* NCTM.

Safir, S., & Dugan, J. (2021). *Street data: A next-generation model for equity, pedagogy, and school transformation.* Corwin.

Teaching for Robust Understanding Framework. (2023). https://truframework.org

Terada, Y. (2018). What's the right amount of homework. *Edutopia.org.* https://www.edutopia.org/article/whats-right-amount-homework

Zavitkovsky, P. (2022). *A question district leaders need to ask more often: What parts of formative assessment can't be outsourced?* Center for Urban Education Leadership. http://www.urbanedleadership.org

CHAPTER 6

Aguirre, J. M., Mayfield-Ingram, K., & Martin, D. B. (2024). *The impact of identity in K-12 mathematics: Rethinking equity-based practices* (2nd ed.). National Council of Teachers of Mathematics.

Chapin, S. H., O'Connor, M. C., & Anderson, N. C. (2009). *Classroom discussions: Using Math Talk to help students learn, grades K-6.* Math Solutions.

Danielson, C. (2016). *Which one doesn't belong? A shapes book, grades K-12.* Steinhouse.

Fletcher, N., & Meador, A. (2022, February). Developing preservice teachers' TPACK through a virtual number talks field experience: A case study. Paper presented at the *Twelfth Congress of the European Society for Research in Mathematics Education (CERME12).*

Hendrickson, S., Hilton, S. C., & Bahr, D. (2008). The Comprehensive Mathematics Framework (CMI): A new lens for examining teaching and learning in the mathematics classroom. *Utah's Mathematics Teacher, 1*(1), 44–52.

Humphreys, C., & Parker, R. (2023). *Making number talks matter: Developing mathematical practices and deepening understanding, grades 3–10.* Routledge.

Jenkins, M. C., & Murawski, W. W. (2023). *Connecting high-leverage practices to student success: Collaboration in inclusive classrooms.* Corwin.

Joswick, C., & Taylor, C. N. (2022). Supporting SEL competencies with number talks. *Mathematics Teacher: Learning and Teaching PK-12, 115*(11), 781–791.

Kazemi, E., & Hintz, A. (2023). *Intentional talk: How to structure and lead productive mathematical discussions.* Routledge.

Lemon, T., & Hendrickson, S. (2023). Building coherence and progression on sound frameworks. *Mathematics Teacher: Learning and Teaching PK-12, 116*(7), 490–502.

National Council of Teachers of Mathematics (NCTM). (2018). *Catalyzing change in high school mathematics: Initiating critical conversations.*

Ramirez, N. (2021, January). Asset-based thinking, language, and actions. Paper presented at the *ASSM Meet-Up for the Association of State Supervisors of Mathematics.*

Rineck, L. M. (2020). *A holistic developmental mathematics course for all learners* (Doctoral dissertation, The University of Wisconsin—Milwaukee).

Seda, P., & Brown, K. (2021). *Choosing to see: A framework for equity in the math classroom.* Dave Burgess Consulting.

Spangler, D. A., & Wanko, J. J. (Eds.). (2017). *Enhancing classroom practice with research behind principles to actions.* National Council of Teachers of Mathematics.

Walter, H. A. (2018). Beyond turn and talk: Creating discourse. *Teaching Children Mathematics, 25*(3), 180–185.

Wilkerson, T. (2021). *Noticing and wondering: Empowerment in learning.* NCTM. https://www.nctm.org/News-and-Calendar/Messages-from-the-President/Archive/Trena-Wilkerson/Noticing-and-Wondering_-Empowerment-in-Learning/

Zwiers, J., Dieckmann, J., Rutherford-Quach, S., Daro, V., Skarin, R., Weiss, S., & Malamut, J. (2017). *Principles for the design of mathematics curricula: Promoting language and content development.* https://ul.stanford.edu/sites/default/files/resource/2021-11/Principles%20for%20the%20Design%20of%20Mathematics%20Curricula_1.pdf

CHAPTER 7

CAST. (2018). *Universal Design for Learning Guidelines version 2.2.* http://udlguidelines.cast.org

Collaborative for Academic, Social, and Emotional Learning (CASEL). (2024). CASEL Framework. https://casel.org/fundamentals-of-sel/what-is-the-casel-framework/#interactive-casel-wheel

Featherstone, H., Crespo, S., Jilk, L. M., Oslund, J. A., Parks, A. N., & Wood, M. B. (2011). *Smarter together! Collaboration and equity in the elementary math classroom.* National Council of Teachers of Mathematics.

National Council of Teachers of Mathematics (NCTM). (2014). *Principles to actions: Ensuring mathematical success for all.*

Open Up Resources. (2021). *Open Up Resources high school mathematics integrated math I.*

Smith, M. S., Bill, V., & Sherin, M. G. (2020). *The Five Practices in Practice [Elementary]: Successfully orchestrating mathematics discussions in your elementary classroom.* Corwin.

Smith, M. S., & Sherin, M. G. (2019). *The Five Practices in Practice [Middle School]: Successfully orchestrating mathematics discussions in your middle school classroom.* Corwin.

Smith, M. S., Steele, M. D., & Sherin, M. G. (2020). *The Five Practices in Practice [High School]: Successfully orchestrating mathematics discussions in your high school classroom.* Corwin.

Smith, M. S., & Stein, M. K. (2018). *5 Practices for orchestrating productive mathematics discussion.* National Council of Teachers of Mathematics.

WIDA. (2020). *2020 WIDA English Language Development Standards kindergarten-grade 12.* Board of Regents of the University of Wisconsin System, on behalf of WIDA. https://wida.wisc.edu/sites/default/files/resource/WIDA-ELD-Standards-Framework-2020.pdf

CHAPTER 8

Larson, M. R., & Kanold, T. D. (2016). *Balancing the equation: A guide to school mathematics for educators and parents (contexts for effective student learning in the common core).* Solution Tree Press.

National Council of Teachers of Mathematics (NCTM). (2014). *Principles to actions: Ensuring mathematical success for all.*

National Council of Teachers of Mathematics (NCTM). (2018). *Catalyzing change in high school mathematics: Initiating critical conversations.*

National Council of Teachers of Mathematics (NCTM). (2020a). *Catalyzing change in middle school mathematics: Initiating critical conversations.*

National Council of Teachers of Mathematics (NCTM). (2020b). *Catalyzing change in elementary and early childhood mathematics: Initiating critical conversations.*

Steele, M. D., & Huhn, C. (2018). *A quiet revolution: One district's story of radical curricular change in high school mathematics.* Information Age.

CHAPTER 9

Candela, A. G., & Boston, M. (2022). Centering professional development around the Instructional Quality Assessment rubrics. *Mathematics Teacher Educator, 10*(3), 204–222.

Charles A. Dana Center, Student Achievement Partners, & Education Strategy Group. (2022). *Re-envisioning Mathematics Pathways to expand opportunities: The landscape of high school to postsecondary course sequences.* https://edstrategy.org/wp-content/uploads/2022/07/Re-Envisioning-Mathematics-Pathways-to-Expand-Opportunities_FINAL.pdf

Collaborative for Academic, Social, and Emotional Learning (CASEL). (2024). *CASEL Framework.* https://casel.org/fundamentals-of-sel/what-is-the-casel-framework/#interactive-casel-wheel

Conference Board of Mathematical Sciences. (2024). *Background to the Pathways Forum.* https://www.cbmsweb.org/cbms_forum_6/background-to-the-forum/

David, J. L. (2007). What research says about pacing guides. *ASCD, 66*(2). https://www.ascd.org/el/articles/pacing-guides

Dufour, R. (2021). *Revisiting professional learning communities at work* (2nd ed.). Solution Tree.

Griffins, B. (2023). *The what, why, and how of learning walks.* Learner Center Collaborative. https://learnercentered.org/blog/learning-walks-what-why-how/

Holcomb, S. (2021). *The history of NEA.* National Education Association.

Individuals with Disabilities Education Act, 20 U.S.C. § 1400 (1990; reauthorized 2004). https://sites.ed.gov/idea/regs/b/d/300.320

National Council of Teachers of Mathematics (NCTM). (2018). *Catalyzing change in high school mathematics: Initiating critical conversations.*

NCTM. (in press). *High school reimagined: Revitalized and relevant.*

National Education Association of the United States. Committee of Ten on Secondary

School Studies. (1894). *Report of the Committee of Ten on Secondary School Studies: With the reports of the conferences arranged by this committee.*

National Governors Association. (2023). *Governors prioritize postsecondary education pathways in 2023 State of the State addresses.* https://www.nga.org/news/commentary/governors-prioritize-postsecondary-education-pathways-in-2023-state-of-the-state-addresses/

Rouleau, K., & Corner, T. (2020). *Classroom walkthroughs: Where data-gathering and relationship building meet for school improvement.* McREL International. https://files.eric.ed.gov/fulltext/ED611283.pdf

Utah State Board of Education. (2024). *Reflective framework for Individualized Education Program development.* https://www.schools.utah.gov/specialeducation/_specialeducation/_speialeducationdirectors/_technicalassistance/DirectorIEPReflectiveFrameworkCompleteManual.pdf

CHAPTER 10

Adiredja, A. P., & Louie, N. (2020). Untangling the web of deficit discourses in mathematics education. *For the Learning of Mathematics, 40*(1), 42–46.

Martin, D. B. (2019). Equity, inclusion, and anti-blackness in mathematics education. *Race Ethnicity and Education, 22*(4), 459–478.

National Council of Supervisors of Mathematics (NCSM). (2014). *It's TIME: Themes and imperatives for mathematics education.* Solution Tree.

Steele, M. D., & Huhn, C. (2018). *A quiet revolution: One district's story of radical curricular change in high school mathematics.* Information Age.

Wiggins, G. P. (1993). *Assessing student performance: Exploring the purpose and limits of testing.* Jossey-Bass.

Wiliam, D. (2003). Formative assessment in mathematics. In L. Haggarty (Ed.), *Aspects of teaching secondary mathematics* (pp. 289–300). Routledge.

Index

**JENNIFER M. BAY-WILLIAMS,
JOHN J. SANGIOVANNI,
ROSALBA SERRANO,
SHERRI MARTINIE,
JENNIFER SUH, C. DAVID WALTERS**

Because fluency is so much more
than basic facts and algorithms.
Grades K–8

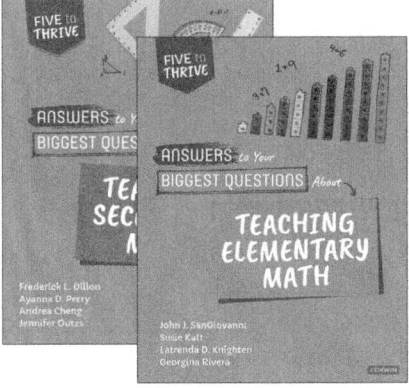

**JOHN J. SANGIOVANNI, SUSIE KATT,
LATRENDA D. KNIGHTEN,
GEORGINA RIVERA,
FREDERICK L. DILLON,
AYANNA D. PERRY,
ANDREA CHENG, JENNIFER OUTZS**

Actionable answers to your most
pressing questions about teaching
elementary and secondary math.

Elementary, Secondary

**ROBERT Q. BERRY III, BASIL M. CONWAY IV,
BRIAN R. LAWLER, JOHN W. STALEY,
COURTNEY KOESTLER, JENNIFER WARD,
MARIA DEL ROSARIO ZAVALA,
TONYA GAU BARTELL, CATHERY YEH,
MATHEW FELTON-KOESTLER,
LATEEFAH ID-DEEN,
MARY CANDACE RAYGOZA,
AMANDA RUIZ, EVA THANHEISER**

Learn to plan instruction that engages
students in mathematics explorations
through age-appropriate and culturally
relevant social justice topics.

**Early Elementary, Upper Elementary,
Middle School, High School**

**SARA DELANO MOORE,
KIMBERLY RIMBEY**

A journey toward making
manipulatives meaningful.
Grades K–3, 4–8

CM22153268

A Sage Company

CORWIN HAS ONE MISSION: to enhance education through intentional professional learning.

We build long-term relationships with our authors, educators, clients, and associations who partner with us to develop and continuously improve the best evidence-based practices that establish and support lifelong learning.